Implementing a Low-Carbon Future

STUDIES IN COMPARATIVE ENERGY AND ENVIRONMENTAL POLITICS

Series editors: Todd A. Eisenstadt, American University, and Joanna I. Lewis, Georgetown University

The Political Economy of Climate Finance Effectiveness in Developing Countries
Carbon Markets, Climate Funds, and the State
Mark Purdon

The Roots of Engagement: Understanding Opposition and Support for Resource Extraction
Moisés Arce, Michael S. Hendricks, and Marc S. Polizzi

Fueling State Capitalism: How Domestic Politics Shapes Foreign Investments of National Oil Companies
Andrew Cheon

Who Speaks for Nature? Indigenous Movements, Public Opinion, and the Petro-State in Ecuador
Todd A. Eisenstadt and Karleen Jones West

A Good Life on a Finite Earth: The Political Economy of Green Growth
Daniel J. Fiorino

Slow Harms and Citizen Action: Environmental Degradation and Policy Change in Latin American Cities
Veronica Herrera

The Politics of Extraction: Territorial Rights, Participatory Institutions, and Conflict in Latin America
Maiah Jaskoski

Democracy in the Woods: Environmental Conservation and Social Justice in India, Tanzania, and Mexico
Prakash Kashwan

Breaking Ground: From Extraction Booms to Mining Bans in Latin America
Rose J. Spalding

Implementing a Low-Carbon Future

Climate Leadership in Chinese Cities

Weila Gong

OXFORD
UNIVERSITY PRESS

OXFORD
UNIVERSITY PRESS

Oxford University Press is a department of the University of Oxford.
It furthers the University's objective of excellence in research, scholarship,
and education by publishing worldwide. Oxford is a registered trade mark of
Oxford University Press in the UK and in certain other countries.

Published in the United States of America by Oxford University Press
198 Madison Avenue, New York, NY 10016, United States of America.

CIP data is on file at the Library of Congress

ISBN 9780197757420

DOI: 10.1093/9780197757451.001.0001

Printed by Marquis Book Printing, Canada

The manufacturer's authorized representative in the EU for product safety is
Oxford University Press España S.A., Parque Empresarial San Fernando de Henares,
Avenida de Castilla, 2 – 28830 Madrid (www.oup.es/en or product.safety@oup.com).
OUP España S.A. also acts as importer into Spain of products made by the manufacturer.

FSC www.fsc.org MIX Paper | Supporting responsible forestry FSC® C103567

致我的家人

For my family

Contents

List of Figures

List of Tables

Acknowledgments

Implementing climate pledges to deliver sustained actions is one of the greatest challenges of our times. This book documents the role of bureaucracy in shaping the divergent paths taken by four Chinese cities in translating low-carbon ideas into policy action in the implementation process. Studying the climate policy implementation process in a centralized political system like China's is extremely challenging. It would have been impossible without the trust and selfless assistance of interviewees who were willing to share their insights with me, a graduate student examining this issue for her dissertation in far-away Germany. While they may be unnamed in this book, it would not have been possible without these bureaucrats, scholars, and policy professionals from inside and outside the government, and from across China and beyond, who took time out of their busy schedules to share their expertise and insights with me.

Working on this book has been a gratifying experience of intellectual exploration and the culmination of my academic journey thus far. I would like to give special thanks to my academic mentors who have shaped my scholarly thinking as I worked on this project, notably Miranda A. Schreurs, Joanna I. Lewis, Henry Lee, and John P. Holdren. My approach to comparative environmental and climate politics and policy has been enormously influenced by working with Miranda, my PhD advisor at the Technical University of Munich. In addition to guiding me as I learned the art of comparison in research, her supportive group of PhD students from all over the world made my doctoral studies a truly enjoyable and enlightening experience. I am grateful for her enduring support, inspiration, and generosity over the years. I substantially revised this book manuscript during my postdoctoral fellowships at Georgetown University's Walsh School of Foreign Service, with Joanna, and at Harvard Kennedy School's Belfer Center for Science and International Affairs, with Henry and John. Joanna's astute advice and sustained encouragement, starting during my PhD and continuing as we collaborated on our projects on the Green Belt and Road Initiative and China's coal transition, have also shaped the book. By engaging closely with my work, Henry and John generously shared their extensive experience at the nexus of academia and policymaking and pushed me to strengthen the project. I am fortunate to have been one of many who have benefited from their lifelong dedication to fostering scholarship and encouraging a younger generation of interdisciplinary scholars committed to addressing climate and environmental challenges around the world.

Several mentors from the world of practice were also instrumental in directing my interests in subnational climate policymaking, particularly at the city level, and in helping me unpack climate policy implementation in China. I would particularly like to thank Zhen Sun at the Ministry of Ecology and Environment's Department of Climate Change, Huaqing Xu at the National Center for Climate Change Strategy and International Cooperation, and Guiyang Zhuang at the Chinese Academy of Social Sciences. Till Kötter was instrumental in opening the door for me to the climate policy community in China after my first, most challenging months in the field. Hangyu Zhao instilled in me an interest in low-carbon energy transitions by guiding me on tours of renewable energy facilities and conferences after I embarked on my PhD studies in Germany. I am also grateful to Jennifer Turner at the Wilson Center for pushing me to explain how my work connects to broader contexts and issues; I thank her for her excellent suggestions.

Many scholars working on climate, environment, politics and policy, and China engaged in productive conversations about my research during the project's evolution in conferences, workshops, and seminars. It is not possible to thank them all individually. In particular, I am especially grateful to Jessica Teets for carefully reading and providing insightful comments and suggestions on several versions of chapters as I finalized the manuscript and her encouragement and support throughout the final stage; Ye Qi for hosting me at the Brookings-Tsinghua Center while I was a PhD student and for providing perceptive guidance on early draft chapters; Thomas G. Rawski for his insights on central–local relations; and Lihua Yang and Kyle Jaros for their productive advice on presenting some of my major findings in article form, which also influenced my approach in the book. My thanks also go to Xun Cao, Jim Cook, Iza Ding, Wenjuan Dong, Min Fan, Thomas B. Gold, David Gordon, Jennifer Hadden, Eugénia da Conceição-Heldt, Qingzhi Huan, Martin Jänicke, John James Kennedy, Daniel Koss, Kevin O'Brien, Xiaoyu Pu, Ran Ran, Rachel Stern, Yixian Sun, Eva Sternfeld, Wenzhao Tao, Alex Wang, Dong Wang, Xinsheng Wang, Xuedong Wang, Bochong Wu, Stefan Wurst, Yunsheng Xie, Yu Xie, Ling Zhang, Li Zheng, and Songli Zhu.

Throughout my academic career, I've made friends in various parts of the world who have supported me with a sense of community and been willing to read or discuss this work during its long journey. Thank you especially to Hamed Beheshti, Hélène Benveniste, Yun Cao, Christopher Carothers, Evelyn Chan, Sau Kau Chan, Kathryn Chelminski, Christine Chen, Natasha

Donevska, Nadia Filimonova, Eyck Freymann, Olivia Gippner, Christine Gschwendtner, Ting Guan, Sacha Hahn, Ruifang Jiang, Vassilca Jeremias, Se Hee Jung, Yi Hyun Kang, Kristine Li, Qing Li, Xinlei Li, Xue Li, Shangwei Liu, Jens Marquardt, Eva Öller, Tianrong Pan, Anurag Panda, Ishana Ratan, Sybil Reitz, Chunhong Sheng, Nhlanhla Sibisi, Kaihui Song, Qijiao Song, and Karoline Steinbacher. I also thank Julia Teebbken, Karen Trebitz, Dongping Wang, Qi Wang, Saul Wilson, Dan Wu, Huaming Wu, Jost Wübbeke, Shiming Yang, Shaojin Zeng, Hong Zhang, Shuo Zhang, and Yun Zhang.

I have received generous funding during the research and writing of this book from the China Scholarship Council, the Environment and Natural Resources Program (ENRP), and the Science, Technology, and Public Policy Program (STPP) at Harvard Kennedy School's Belfer Center for Science and International Affairs. I also acknowledge institutional support from the Freie Universität Berlin's Graduate School of Global Politics and Environmental Policy Research Center, the Technical University of Munich's Graduate School, the Brookings-Tsinghua Center, the University of Pittsburgh's Asian Studies Center, and the Science, Technology and International Affairs Program (STIA) at Georgetown University's Walsh School of Foreign Service, the UC Berkeley School of Law, and the UC San Diego's School of Global Policy and Strategy's 21 ST Century China Center.

I owe thanks to the wonderful communications specialists who supported my work, including Kelley Friel, for her careful edits, which greatly improved this manuscript; D. Scott, for their valuable structural advice; Pia Kohler, for her insightful and precise editorial inputs; and Jessica Wallace for copyediting support in the early days of this research project. Elizabeth Hanlon, for her advice on the manuscript and on communicating my research to a broader audience.

At Oxford University Press, I wish to thank the co-editors of the book series on Studies in Comparative Energy and Environmental Politics, Todd A. Eisenstadt and Joanna I. Lewis; Angela Chnapko, for supporting this undertaking; and Pheobe Aldridge-Turner, for help throughout the production process. I also thank the two anonymous reviewers for their constructive comments and suggestions.

I am grateful to family and friends who generously hosted me, including Chunhong Sheng, Xiong Xiao, Ping Yü, and Huifang Chen for offering me a place to call home during my multiple field trips to China over many years.

Finally, this book is dedicated to my family, whose love, patience, and support made it possible. I often thought of my grandma, Xiong Wu, who gracefully embraced her life with independent thinking, resilience, and gratitude and set an example that has become a source of strength that I have continuously turned to while completing this work. I am deeply grateful to my parents, Yingjian Gong and Min Xiao, who instilled in me the belief that I could pursue my passion to live my life to the fullest while they will always have my back. My aunt Jian Xiao and my cousins Jingsong Hong and Qiao-qiao Zhou often grounded me with their worldly wisdom and laughter even though we live thousands of miles apart. I appreciate the steadfast support of my husband, Lei, and two daughters, Nora and Charlotte—they bring light and joy into my everyday life.

Weila Gong
Berkeley, California

1
Introduction

As the world's largest carbon emitter and a key investor in clean energy, China plays a particularly important role in shaping the direction of global efforts to address climate change. It has pledged to peak its carbon emissions before 2030 and to achieve carbon neutrality before 2060. While China led global renewable energy investment between 2010 and 2023 (REN21 2011; REN21 2024), it has struggled to reconcile the competing priorities of economic growth, environmental protection, and decarbonization (Xue et al. 2007). Although subnational governments have a reputation for prioritizing economic interests over environmental protection, they have also helped China become a world leader in renewable energy technology manufacturing (Nahm 2017). Given the country's vast size and significant socioeconomic diversity across regions, the central government relies heavily on subnational governments to implement national polices at the local level. Therefore, the extent to which China can fulfill its climate pledges largely depends on subnational implementation.

This book investigates *subnational climate leadership*, focusing on how municipal governments translate low-carbon goals into implementable policy actions, using the case of China's low-carbon city experimentation. Policy experimentation in China is a unique process in which the national government grants subnational governments considerable leeway to test new policy ideas and mechanisms before rolling them out nationwide (Schreurs 2008). The National Development and Reform Commission, the leading government agency responsible for the country's national macroeconomic planning and energy issues, launched the national low-carbon pilot city program examined in this book in 2010. It was the first nationwide climate policy experimentation program since climate change emerged as a distinct policy area from environmental protection in 2008 (Teng and Wang 2021). Since then, addressing climate change has been translated as low-carbon development in local areas and regions; the emphasis has been on decoupling growth from increased fossil fuel energy consumption (Zhou et al. 2012). The experimental program had no obvious policy or bureaucratic incentives, such as national financial subsidies or targets in the bureaucratic evaluation system, which the national government would normally use to encourage

Implementing a Low-Carbon Future. Weila Gong, Oxford University Press. © Oxford University Press (2025).
DOI: 10.1093/9780197757451.003.0001

subnational participation. Yet many cities have voluntarily participated in the program and have demonstrated different levels of policy engagement over time, resulting in distinct patterns of uneven subnational policy engagement.

The book addresses two central questions. First, why are some local areas and regions doing better than others at initiating and maintaining low-carbon policy experimentation? Second, how do local governments overcome the political and financial constraints inherent in the centralized political system to implement low-carbon policy experiments?

Explaining Subnational Engagement in Low-Carbon Policy Experimentation

Local areas and regions have been a powerhouse for low-carbon policy experiments. Subnational (i.e., provincial-, municipal-, and township-level) governments have long engaged in policy experimentation, which has informed national responses to decarbonization. In the 1970s, remote rural areas were the first to develop domestic renewable projects such as small-scale hydropower and mini-wind turbines. In the 1980s, local governments, under the auspices of national experimental initiatives, began to expand non-fossil-fuel energy sources to address the energy scarcity in rural areas. While the Chinese government started to participate in international climate negotiation in the 1980s, climate change only began to gain political significance linked to economic development on China's domestic policy agenda in the late 1990s (Lewis 2012, 12). Since then, subnational engagement in international climate projects, such as the Clean Development Mechanism (CDM), has substantially shaped the national response to reducing carbon emissions. For example, the growth in subnational participation in the CDM—which yielded substantial economic benefits for local renewable energy projects—made the national government more responsive to the operation of international climate mechanisms; China hosted the most CDM projects in the world in 2007 (Heggelund 2007; Schröder 2012, 2). Afterwards, a growing number of provincial and municipal governments began exploring ways to reduce carbon emissions across local industrial, building, and transportation sectors, starting with the national low-carbon pilot city program.

A common shortcoming of climate policy experiments is that they are hard to maintain long enough to be institutionalized into the existing policy system. Like experimental programs in other countries, subnational policy experiments in China are mostly voluntary and are often hard to sustain, particularly after changes in political leadership (Fewsmith 2013, 66; Sabel and Victor 2022, 3). For example, in a centralized political system like China's, top political leaders (such as mayors and party secretaries in municipal governments) determine the priorities of local government initiatives and whether there is sufficient political willingness to engage in policy experimentation. Yet these top leaders are frequently reassigned to other cities; the majority of them stay less than five years in each position on average (Kou and Tsai 2014; Eaton and Kostka 2014). Thus, if formal institutions (e.g., regulations, standards, and legislation that are central to influencing behavioral changes at the individual and organizational levels) have not been created to sustain the momentum of the policy experimentation, it is unlikely to be successful after a turnover in government leadership. A new mayor with different aims could squander these achievements by shifting political attention to new policy priorities. Policy experimentation in China has thus been criticized for being vulnerable to political turnovers, particularly due to its lack of institutionalization (Fewsmith 2013, 155).

According to data from the Ministry of Ecology and Environment (Ministry of Ecology and Environment 2023b), as of 2023, forty percent of the cities that participated in the national low-carbon pilot city program had started to compile greenhouse gas (GHG) inventories, but only six percent had created guidance or standards for carbon accounting. While thirty-four of the eighty-one pilot cities had issued work plans for low-carbon city development by this time, only four (Shijiazhuang, Nanchang, Tianjin, and Shenzhen) had passed local legislation on regulating carbon emissions (Ministry of Ecology and Environment 2023b).

Prior research on China's low-carbon energy transitions has assessed the technological and policy pathways that are crucial for realizing the country's climate pledges, but it has been less explicit about the politics surrounding the policy implementation process. Previous studies of the pathways leading to China's attainment of its carbon neutrality goals tend to concentrate on technological solutions and economic incentives (Lee et al. 2021; Liu et al. 2022). Research on subnational low-carbon policy engagement has often been limited to the early stage of the policy process, such as agenda setting and policy formation (Filimonova 2024; Lo et al. 2015; Peng and Bai 2018; Qi et al. 2008; Song et al. 2021; Tie et al. 2020). However, less is known about the varying local policy engagement in initiating and sustaining low-carbon policy experiments—particularly the essential role of local bureaucrats in the mobilization process. Green and Reyes (2023) have called for more in-depth research on the policy implementation process; although hundreds of countries and cities have declared their intention to go carbon neutral, they note "an urgent need for greater understanding about the political conditions which promote or constrain the creation and implementation of net zero policies across sectors."

Why have some local areas and regions engaged more substantially than others (and how)? Why are some Chinese cities able to initiate and maintain their low-carbon policy engagement, while others are not? Four arguments can be put forward to account for such different outcomes, but each fails to adequately explain what is happening. First, some scholars assert that the government's environmental policies have been influenced by the economic capacity and industrial structures in their jurisdictions (Grossman and Krueger 1995; Vogel 1997). While economic factors may influence local engagement in low-carbon policy experiments, they cannot fully explain the policy changes enacted in particular cities, including less affluent ones. Indeed, contrary to the expectations of conventional wisdom—which maintains that economic development status is the primary factor influencing governments' attempts to address environmental challenges—this book demonstrates that some less affluent pilot cities have more actively engaged in low-carbon initiatives than more prosperous ones.

A second argument relies on principal–agent theory in the form of China's central–local relations: local leaders' responsiveness to central political appeals is influenced by the gains they expect to receive in terms of bureaucratic incentives or political patronage. Principal–agent theory makes an important contribution to this topic, but it has a limited ability to explain local engagement in an issue area *not* included in bureaucratic evaluation metrics or where central government connections do not promise potential benefits to local leaders, such as the pilot program this book evaluates. Strong central political patronage may indeed intensify local engagement in policy experimentation, as it may insulate local leaders from the negative repercussions of failed policy experiments (Chung 2000, 9–10). Yet ministry-level officials (rather than top political leaders in Beijing) introduced and coordinated the national low-carbon pilot city program examined in this book; such officials lack sufficient political authority to shield local officials from the perceived risks of failed policy experiments. Therefore, the principal–agent perspective cannot adequately explain subnational engagement in low-carbon policy experimentation that does not produce immediate political and economic gains for local government leaders.

The third explanation refers to international influence through participation in transnational city networks on climate change. Cities all over the world tend to defer to international authority for political and financial benefits as they participate in transnational climate change networks (Gordon 2020, 219). However, the central government has prevented Chinese cities from actively engaging in such networks (Gordon 2020, 229). International experts and agencies helped each of the four case cities shop policy ideas and frame the issue, but they were not involved in local policy implementation. Therefore, domestic politics, rather than transnational networks or governance, drive the implementation process.

The final alternative argument relates to the role of business actors in climate politics and policymaking. Interest groups in democratic states influence subnational government climate engagement through lobbying (Culhane et al. 2021; Stokes 2020, 7). Unlike interest groups that can wield political influence through the party, courts, or public opinion (Stokes 2020, 7), businesses in China are unlikely to directly influence government policymaking and policy implementation until they form a coalition with bureaucrats (Sun 2022, 145). Local bureaucrats can influence local businesses' incentives and behaviors (Chen 2018, 158). According to an interviewee who participated in the launch of a local emissions trading system, local enterprises were skeptical of the system when they were asked to join; they worried it was a new form of government rent-seeking (Interview 240107a 2013). This suggests that local governments, rather than business actors, are driving the process.

This book argues that economic development status, central–local relations, and external incentives (including international city networks and business lobbying) cannot adequately explain the variation documented in the subnational engagement in China's low-carbon city experimentation. This is largely because prior research often overlooks the implementation process of how local government leadership translates political pledges into policy actions during the experimentation. The book proposes an approach to examining uneven subnational engagement in low-carbon policy implementation through the lens of subnational government leadership.

A Leadership Approach to Examining Uneven Subnational Climate Engagement

Conventional theoretical lenses emphasize how external influences such as central–local relations or bureaucratic incentives and evaluation criteria shape subnational policy engagement on environmental issues in China (Cao et al. 2019; Ding 2020; Ran 2013; Shen 2022; van der Kamp 2023). These perspectives often downplay the role of local government leadership, particularly the essential role of bureaucratic entrepreneurs, in translating policy goals into sustained action in implementation.

My leadership approach is instead situated within process thinking: I treat subnational low-carbon policy engagement as a *local mobilization process*. I argue that understanding the organizational structure of local government leadership and its resource mobilization strategies can be as important as appreciating the external incentives to which leaders are responding in the implementation process.

My framework merges the key elements of the political leadership, policy process, and resource mobilization literatures to explain how local government leadership translates policy ideas into actions. Political leadership research is concerned with what Bryan D. Jones (1989, 4) calls the problem of agency–structure interaction; Jones inquires, "To what extent are the actions of leaders determined (or put in a weaker form, constrained) by forces beyond the leader's control? To what extent is leadership dictated by structure, and to what extent is there room for independent action?" Scholars studying leadership and resource mobilization assert that it is not only individual leaders but also the *leadership's organizational structure and networks* that affect resource mobilization strategies and the effectiveness of leadership capacity (Ganz 2000; Morris and Staggenborg 2007; Selznick 1984). Recognizing that agencies operate within myriad layers of institutional structures suggests that they have the potential to form alliances within the existing structures or create new institutional structures to mobilize the necessary resources to implement sustainable policies.

Other scholars examining leadership and innovation in the government such as James W. Doig and Erwin C. Hargrove (1990, 7–17) point out the importance of *entrepreneurial executives* "whose careers at managerial levels were linked to innovative ideas and efforts to carry these ideas into effect" to influence sustained attempts to affect their organizations through goal setting and implementation. In this view, sustained government efforts emerge from entrepreneurial bureaucrats and their leadership strategies, rather than top political leaders.

This perspective reveals the limitations of conflating local government leadership with local political leaders when examining policy experimentation. While the conventional approach to studying government leadership emphasizes the importance of local political leaders, it tends to underplay the role of local *bureaucrats* in the process of sustaining policy actions during implementation. In most cases, political leaders are agenda setters, but bureaucrats have a substantial impact on the process of implementation and policy formation (Kingdon 1995, 31). Insights from policy process theory highlight the crucial role of policy entrepreneurs in identifying political windows and building alliances to push new issues onto governments' policy agendas (Kingdon 1995, 168–170). Rabe (2004, 24) articulates that "entrepreneurs must begin to build bridges with the help of their allies, to legislators and governors—and their aides—who can be persuaded of the political benefits of supporting such steps." Indeed, even working within the given political and resource constraints, local bureaucrats can act as policy entrepreneurs to build alliances with key actors inside and outside the government to mobilize wider support for low-carbon policies tailored to their jurisdictions.

My conception of Bridging Leadership builds on but goes beyond these approaches to study governments' (and organizations') decision making in uncertain situations by emphasizing the bridging role of bureaucratic entrepreneurs and their mobilization strategies in *initiating and sustaining* policy actions throughout the implementation process. By explaining how local bureaucrats leverage their government positions to employ mobilization strategies for policy implementation, my leadership approach showcases the agency–structure interactions behind subnational climate leadership in translating low-carbon pledges into actions. It also explains the importance of subnational government leadership and the local policy process in influencing the divergent outcomes of policy experiments in a centralized political system like China's.

What Is Bridging Leadership?

I define *bridging leadership* as the process employed by bridge leaders, who mobilize the necessary political support and implementation alliances to translate proposed policy ideas into actions. As depicted in Figure 1.1, *bureaucratic entrepreneurs*, who are often mid-level bureaucratic leaders (at the deputy *chu* level and above, including local bureaucrats ranging from department heads up to vice mayors) in China, are the *bridge leaders* (the spokes). While top *political leaders* (in the hub), such as mayors and party secretaries, rotate in and out of senior posts in different localities or are promoted to higher-level governments, mid-level bureaucrats often spend their entire careers in a single municipality (Ang 2016, 108; Kennedy and Chen 2014). This allows them to rise to decision-making positions and command administrative resources within their respective bureaucratic systems for specific policy tasks. These bureaucratic leaders also gain specialized expertise and insights into local

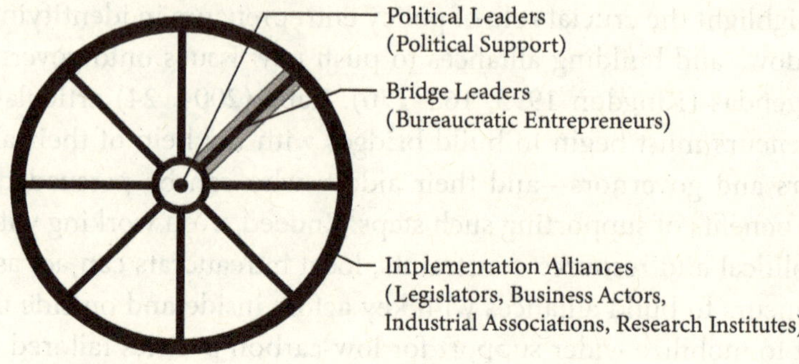

Figure 1.1 The Organizational Structure of Bridging Leadership
Source: Author's illustration.

political dynamics through their involvement in the local policymaking process.

These mid-level bureaucrats operate as a bridging bureaucratic tier. They prepare policy drafts, provide counsel to political leaders, and review, approve, and supervise local development projects. They act as implementation bridges to garner support from political leaders and build *implementation alliances* (the outer rim) with key local actors such as legislators, business actors, industrial associations, and research institutes to drive low-carbon policy implementation.

Drawing on four comparative case studies, process tracing, and over one hundred in-depth interviews, this book highlights how the availability of bridge leaders and their mobilization strategies have generated different levels of local engagement in low-carbon policy experimentation. Pilot cities in which bridge leaders secure local politicians' support and build alliances with key government and non-government actors during the policy process are more likely to create an active low-carbon policy program and bring about sustained implementation.

The Mechanism of Bridging Leadership

How do bridge leaders translate policy pledges into actions in low-carbon policy experimentation? In this section, I first elucidate the bridging leadership mechanism that captures the mobilization process which leads to varying policy implementation outcomes (Figure 1.2). I then discuss the key strategies that bridge leaders often employ to mobilize subnational low-carbon policy implementation.

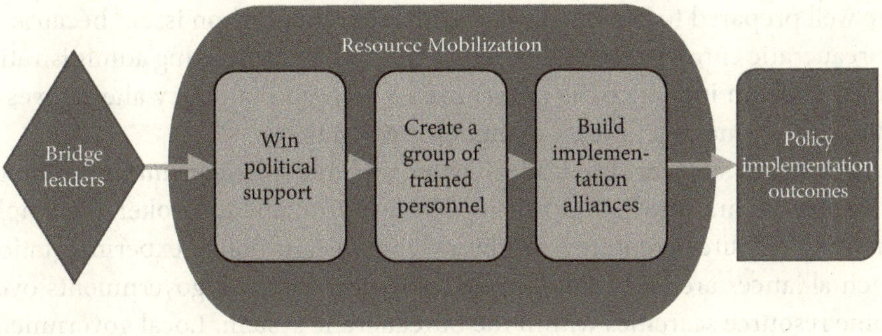

Figure 1.2 The Process of Bridging Leadership in Mobilizing Climate Policy Implementation
Source: Author's illustration.

As illustrated in the middle of Figure 1.2, it is important for bridge leaders to win *political support* from political leaders to help push the issue onto the government agenda, particularly by persuading them of the benefits of acting in the issue area. More importantly, bridge leaders can engage political leaders in the policy process to mobilize wider support for policy implementation. In Chinese cities, party secretaries and mayors are the top local political leaders: they can influence almost every key decision, from setting policy priorities to supervising their implementation (Eaton and Kostka 2014). Securing their support can elevate an issue's priority on the local policy agenda and help overcome cross-divisional collaboration challenges within the government, which are a key barrier to policy implementation in China's fragmented bureaucratic system (Lieberthal and Oksenberg 1988, 22–23).

Next, bridge leaders need to create *a group of trained personnel* who are qualified to implement low-carbon policies to assist local capacity building and policy implementation. These specialists can be an indispensable resource to assist the implementation of climate initiatives from the proposal stage to completion. Bridge leaders' efforts to create such a group can be formal, such as establishing a low-carbon government agency with a full-time staff, or informal, such as commanding a research group under their immediate supervision consisting of local officials and experts from local universities or research institutes to draft policy proposals or action plans. These officials become specialists by being exposed to low-carbon and climate issues through their participation in these formal and informal venues. They become familiar with the issue area through "learning by doing," as well as participating in training programs that have often been jointly organized by the national government and international agencies.[1] These trained officials are well prepared to become bridge leaders for low-carbon issues because, as bureaucratic entrepreneurs, they gain expertise by completing administrative tasks, and can influence the government's selection of policy alternatives in the decision-making process (Kingdon 1995, 30–34).

In a third step, bridge leaders need to build *implementation alliances* with key actors inside and outside the government to broker the implementing resources required to initiate and sustain policy experimentation. Such alliances are essential because they can help local governments overcome resource scarcities within the bureaucratic system. Local governments can use them not only to redistribute fiscal and human resources to prioritized actions within their own bureaucratic system, but also to broker support from key actors in higher levels of government or from local legislative bodies. Given their strategic position and their connections to top political leaders and key actors—such as legislators, and industrial and business sectors—inside and outside the government, bridge leaders are well positioned to help build formal institutions or less formal institutional structures to sustain policy experiments in a new issue area. Baumgartner and Jones (2009, 101–102) similarly find that policy advocates in Western countries can build substantial alliances, particularly in an emerging issue area with few opponents, by aligning with those who support and foster a policy program.

Understanding the process of local bridging leadership in initiating and sustaining climate actions helps unpack the uneven subnational engagement in China's low-carbon policy experimentation. This book shifts the research focus from political leaders' role in shaping local agendas to that of mid-level bureaucrats in orchestrating the implementation of local policy experiments. I go beyond the top-down tournament style of policy implementation (Zhou 2008), which focuses on how national-level bureaucratic evaluation criteria drive local government-led initiatives and assumes that all Chinese local governments react similarly to central appeals. My documentation and analysis of subnational policymaking highlight the importance of local mid-level bureaucrats and their resource mobilization strategies in shaping the uneven low-carbon policy outcomes—which affect China's ability to achieve its climate goals.

Key Arguments of the Book

This book is organized around three key arguments. Together, they help explain local variation in low-carbon policy engagement, how some areas can maintain their engagement while others cannot, and how that affects their subnational climate policy implementation.

First, while climate pledges from Beijing matter, local climate policy engagement varies across and within localities over time, which makes bridging leadership important. The outcomes of local low-carbon policy experiments are largely determined by the availability of local bridge leaders and their resource mobilization capacity rather than by central government commands. Jurisdictions that have local bridge leaders (typically mid-level bureaucrats) who are able to (1) win the support of top political leaders by persuading them of the political benefits of acting on decarbonization, (2) create a group of trained personnel who are qualified to implement low-carbon policies, and (3) build implementation alliances (primarily with key local actors, as well as others through multi-layered governing structures) are better able to maintain low-carbon policy engagement than areas without such bureaucrats.

In a second key argument, in contrast to the conventional wisdom that criticizes policy experiments as being hard to sustain and particularly vulnerable to political turnovers, I argue that bridging leadership can sustain policy experiments during changes in political leadership. While top leaders who regularly rotate posts have a limited interest in creating formal institutions out of the experiments (Fewsmith 2013, 6), the mid-level bureaucratic entrepreneurs who run them are uniquely positioned to introduce institutional efforts to defend their growing bureaucratic interests in the new issue area. As Ikenberry (1988, 192) notes, "The state is not simply an area of political conflict; it is an institutional structure that gives government officials a unique standing in the pursuit of their own goals and strategic perspectives." Bridge leaders typically operate in a brokering bureaucratic tier: by providing counsel to political leaders and reviewing project developments, they can gain political support and insights into the structures of policy opportunities to which they would otherwise not have access. Because institutional structures connect the actors involved and channel the resources needed to complete specific organizational tasks (Manion 2015), each city's ability to institutionalize low-carbon policies and to create new climate institutions helps determine the outcome of its policy experimentation.

Third, it is important for local governments to facilitate the process of low-carbon institutionalization during policy experiments, particularly by pairing new climate institutions with a specialized staff to oversee policy implementation. Even with established formal institutions, such as regulations and standards, policies cannot be enforced without trained specialists, particularly when the issue area is relatively new. China does not lack environmental laws or standards, but reliable *enforcement* (van Rooij 2006; Wang 2018). While mid-level bureaucrats rarely move out of their locality, they are regularly promoted to other positions or retire, and they are sometimes removed from their posts due to rent-seeking behaviors.[2] Local areas without a dedicated division of full-time staff overseeing implementation can struggle with enforcement after bridge leaders change roles. Institutionalizing low-carbon policy with a group of full-time specialists overseeing low-carbon policy implementation is essential to establish and expand low-carbon interests within the existing policy system.

The importance of local bridging leadership cannot be overstated because the national climate pledges cannot be met without subnational policy implementation. Political support and implementation alliances—often mobilized by bridge leaders—influence policy outcomes. This book illustrates three ways in which bridging leadership shapes subnational climate engagement. First, once bridge leaders were able to broker political support and implementation alliances on low-carbon issues, some local governments began to introduce institutional efforts—such as legislation, regulations, and dedicated bureaucratic agencies staffed by specialists—toward decarbonization. Second, the lessons from city-level policy experimentation can scale up to higher levels of government when the bridging leadership is able to extend the implementation alliances upwards. Third, as a result of the bridging leadership's mobilization to institutionalize low-carbon policy during the experimentation, local areas and regions were able to sustain local low-carbon initiatives despite changes in political leadership and even bridge leaders leaving.

I do not argue that subnational climate leadership is a silver bullet. Local governments still have much work to do before their institutional foundations are strong enough to achieve their decarbonization goals. But local bureaucrats have influenced the climate policymaking process in important ways by crafting new climate institutions and specialized enforcement teams that are essential for sustained low-carbon policy implementation. Examining the process of subnational low-carbon policy experimentation suggests ways in which local governments may mobilize the implementation process to make climate policy durable.

Study Design and Data Collection

This book examines the role of subnational climate leadership in China's climate governance by comparing the approaches of four pilot cities

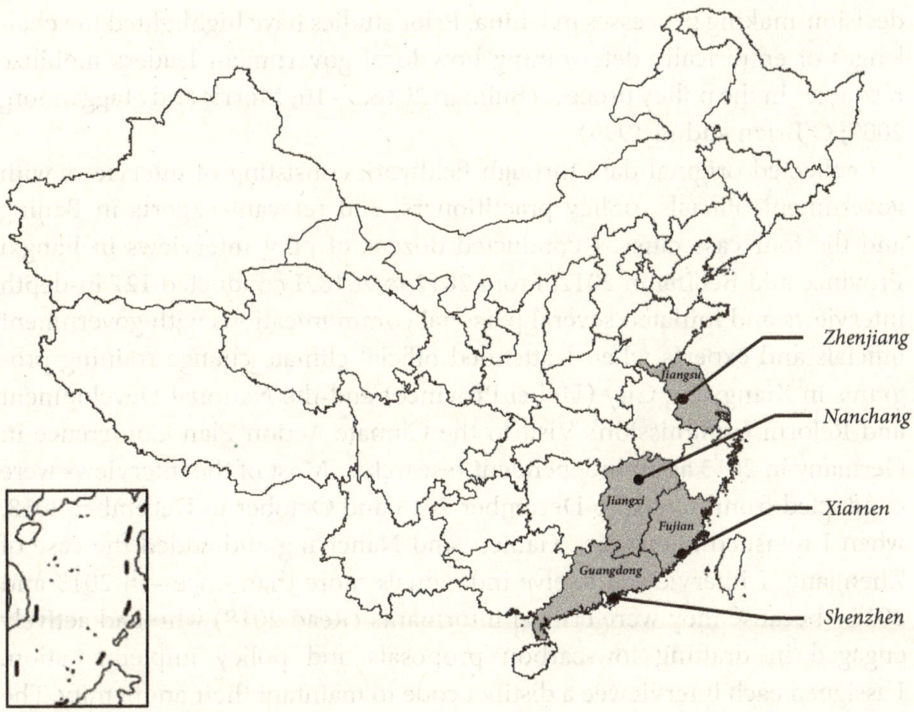

Figure 1.3 Map of the Case Cities
Source: Author's illustration.

(see Figure 1.3)—Shenzhen, Nanchang, Zhenjiang, and Xiamen—to initiating and maintaining their low-carbon policy experiments. They joined the national low-carbon pilot city program at roughly the same time, when the national government was offering few bureaucratic and financial incentives, and share similar goals: to decarbonize local development and reduce reliance on natural resources and fossil fuels. Since becoming pilot cities, all four have set up a leading group (chaired by a mayor or vice mayor) to coordinate the implementation process and proposed low-carbon targets and related implementation measures in the industrial, building, and transportation sectors. Over time, however, they developed different levels of policy institutionalization as a result of their engagement in the policy experiments. Some of the pilot cities created new low-carbon institutions consisting of standards, regulations, and legislation for decarbonization, even though not all of them had a dedicated full-time staff to enforce policies.

Documenting subnational governments' responses to low-carbon policy experimentation and their resource mobilization strategies requires overcoming the difficulties associated with collecting data on government decision-making processes in China. Prior studies have highlighted the challenges of empirically determining how local government leaders mobilize resources in the policy process (Bulman 2016, 7–10; Morris and Staggenborg 2007; O'Brien and Li 1999).

I collected original data through fieldwork consisting of interviews with government officials, policy practitioners, and relevant experts in Beijing and the four case cities. I conducted dozens of pilot interviews in Jiangxi Province and Beijing in 2012. From 2012 to 2018, I conducted 127 in-depth interviews and initiated several personal communications with government officials and experts when I attended official climate change training programs in Xiangyang City (Hubei Province) and the National Development and Reform Commission's Visit to the Climate Action Plan Conference in Germany in 2013 as an independent researcher. Most of the interviews were conducted from August to December 2013 and October to December 2018, when I revisited Shenzhen, Xiamen, and Nanchang and added the case of Zhenjiang. I interviewed twelve individuals more than once—in 2013 and 2018—because they were critical informants (Read 2018) who had actively engaged in drafting low-carbon proposals and policy implementation. I assigned each interviewee a distinct code to maintain their anonymity. The case study analysis was enriched by materials from local government documents and work plans, annual evaluative reports, and low-carbon project development portfolios.

Overall, I find that whether cities can institutionalize low-carbon policy as a result of the experimentation largely hinges on the extent to which local bridge leaders can secure the necessary political support and implementation alliances. Figure 1.4 categorizes the four types of policy engagement on decarbonization based on high and low levels of both aspects in the representative case cities. As the empirical case chapters illustrate, Shenzhen

		Political Support	
		High	Low
Implementation Alliances	High	Substantive Engagement (Chapter 4: Shenzhen)	Performative Engagement (Chapter 5: Nanchang)
	Low	Symbolic Engagement (Chapter 6: Zhenjiang)	Sporadic Engagement (Chapter 7: Xiamen)

Figure 1.4 Types of Local Engagement in Policy Experimentation

Source: Adapted from Gong (2022). The names of the categories were inspired by Ding (2020), Ding and van der Kamp (2020), and Wang (2018).

demonstrated *substantial* engagement with local legislation, regulations, and standards (particularly for its carbon emissions trading system) and assigned full-time staff members to work on implementing the policies. Nanchang *performatively* engaged: its policy experiments were mainly driven by a desire to impress the national government to gain recognition and resources; it did not hire specialists to enforce policy implementation. Yet even for policy experiments related to advanced institutional efforts such as scaling up low-carbon city standards to upper levels of government, as Zhenjiang did as part of its *symbolic* engagement, policy enforcement is not always guaranteed, particularly due to limited industrial participation. Xiamen took a cautious approach of *sporadic* engagement and introduced few institutional efforts during its policy experimentation.

Plan of the Book

Chapter 2 examines the role of subnational climate initiatives in the broader political context of China's low-carbon energy transitions. The chapter explains how such initiatives have influenced national engagement in addressing climate change through policy experimentation. Many observers have claimed that factors such as enhanced capacity or local competition for economic opportunities are key to explaining China's decarbonization and have described the country's low-carbon energy transition as a top-down initiative. I offer a new perspective that focuses on the role of subnational government leadership in shaping the national approach to low-carbon energy transitions.

Past research focuses largely on factors related to central–local relations and the fragmented bureaucratic structure to explain subnational government responses to decarbonization in China. In Chapter 3, I elaborate my alternative theoretical framework—*Bridging Leadership*—designed to help understand the uneven local climate policy engagement, which is based on the role of bridge leaders and the mobilization strategies they pursue in local policy processes.

The following four chapters examine the case studies in detail. Chapter 4 evaluates Shenzhen, a low-carbon pioneer. It was China's first special economic zone and remains a leader in efforts to reduce energy intensity. It has established a degree of low-carbon institutionalization consisting of legislation, regulations, and standards together with a group of trained personnel through its climate policy experimentation. The chapter traces the city's engagement in two major low-carbon experimental programs—the

Shenzhen emissions trading pilot program and the Shenzhen International Low-Carbon City Program. I argue that its bridge leader's ability to secure political support and mobilize alliances with key local actors, such as the local people's congress and business sectors, has enabled the city to create formal institutions and a group of specialists to enhance policy implementation that led to its *substantive engagement* in low-carbon energy transitions.

Chapter 5 turns to Zhenjiang. Its commitment to addressing climate change was unexpected due to its carbon-intensive energy structure and because its coal-powered plants supply electricity to eastern China. I explore how Zhenjiang has displayed *symbolic engagement* in decarbonization. I highlight its unresolved challenges in balancing economic development and reducing its carbon emissions. For example, while bureaucratic entrepreneurs in the city have created one of the first local bureaucratic agencies supervising low-carbon development and pushed its local low-carbon city evaluation standards to the provincial level, insufficient industrial participation has jeopardized local policy implementation.

Chapter 6 analyzes the case of Nanchang. Although it was one of the first Chinese cities to pass local legislation for low-carbon development, Nanchang struggled to maintain its low-carbon policy initiatives. I use this case to illustrate the critical role of mid-level bureaucrats in mobilizing financial and political resources, facilitating new institutions, and creating implementation alliances to overcome the resource constraints—even in an economically lagging inland city—for low-carbon development. Nevertheless, I categorize Nanchang as participating in *performative engagement* because its policy experimentation has been driven by the bridge leader's desire to meet expectations from higher levels of government to access financial resources and obtain personal benefits; enforceable implementation outcomes have largely been ignored.

Chapter 7 traces Xiamen's low-carbon initiatives. Although it is a leader in environmental protection, the city has initiated fewer institutional efforts for decarbonization than the other cases I examine. Its proposed program to become a low-carbon city was not fully implemented. I argue that Xiamen's *sporadic engagement* was largely due to the lack of bridging leadership.

Chapter 8 briefly summarizes the book's key contributions and reflects on broader questions including the opportunities, challenges, and future of subnational climate leadership in China. I also suggest several ways in which China can strengthen subnational engagement in addressing climate change. I conclude the book by reflecting on the limits of subnational climate leadership and the role of subnational climate diplomacy in the era of global geopolitical tensions.

Conventional wisdom often depicts policy experiments as mere political spectacles that rarely achieve their proposed goals. While the fragmented government structure and bureaucratic competition are key to explaining subnational responses to central policy appeals in China, I argue that both characteristics overlook the mobilization process in local policy implementation. This book develops a process-oriented approach that merges key elements of the theories of political leadership, policy process, and resource mobilization to explain the essential role of local bureaucratic entrepreneurs and their mobilization strategies to shape the direction of local engagement in policy experimentation.

The book establishes that bridging leadership in Chinese cities can be a useful theoretical framework through which to study uneven subnational climate policy implementation. Breaking with the idea that China's climate action has been a predominantly top-down process, I highlight the importance of local policy processes and mid-level bureaucrats, who have a crucial role in facilitating the creation of new climate institutions and developing staff capacity to maintain the momentum necessary for low-carbon energy transitions. This can further influence the national approach to addressing climate challenges.

Implementing climate policy in an era of intensified geopolitical tensions, particularly between China and major world players such as the United States, can be challenging. My study clarifies that the Chinese national government can benefit from paying more attention to its subnational climate leadership. Beijing must give subnational governments sufficient leeway to continue experimenting with policy solutions, especially when addressing policy challenges that transcend national boundaries, such as climate change. The Chinese government should further involve non-governmental actors in its climate policy decision making and implementation and encourage non-state actors—such as universities, think tanks, and non-governmental organizations—to take on a bridging leadership role. As part of the process of low-carbon policy institutionalization, the creation of low-carbon institutions and trained staff outside the government can help local areas and regions sustain their climate engagement during changes in local government leadership and navigate the complex global geopolitical situation. Non-state actors' underexplored ability to bring about policy change must also be considered when examining the implementation of subnational low-carbon energy transitions. The lessons learned from this book can be useful for those interested in understanding the role of subnational climate leadership in helping China attain its 2060 carbon neutrality goal and contribute to global efforts to address climate change.

2
Subnational Engagement in China's Low-Carbon Energy Transitions

This chapter explores the development of China's subnational low-carbon actions, which can be traced back to the introduction of energy conservation and renewable energy policy programs in the late 1970s and 1980s. Until the 2000s, subnational governments were mainly perceived as implementers of central directives to improve energy efficiency and develop renewable energy sources. With the central government's ever-increasing international engagement on climate change and the integration of the issue of climate change into national strategy, local areas and regions have become powerhouses of low-carbon policy experimentation.

Subnational low-carbon engagement has greatly influenced China's path toward decarbonization. Subnational governments (those at the provincial, municipal/prefectural, and township levels) are expected to implement central policy directives, but Beijing also delegates considerable authority to local governments to adapt these policies to local conditions (Florini et al. 2012). Growing subnational low-carbon policy experimentation has transformed China's low-carbon initiatives from piloting renewable technologies in rural and remote areas to becoming a world leader in clean technology manufacturing (solar photovoltaic [PV] systems, wind systems, and batteries) as of 2023 (International Energy Agency 2023). This engagement has mainly helped alleviate energy poverty, develop renewable energy industries, and facilitate low-carbon policy experimentation. Climate change emerged as a distinct policy area in the late 2000s. Unclear governing structures, insufficient staffing, and frequent institutional reshuffling have limited the potential for subnational engagement to address climate change. Yet, such engagement has proven to be instrumental in shaping the country's approach to decarbonization in recent decades; its potential growth has not abated as China approaches the deadlines imposed by its carbon peak and neutrality pledges.

This chapter links the history of subnational engagement to China's paths toward low-carbon energy development and evolving responses to climate

Implementing a Low-Carbon Future. Weila Gong, Oxford University Press. © Oxford University Press (2025).
DOI: 10.1093/9780197757451.003.0002

change in recent decades. I show that such engagement has shaped the country's approach to addressing climate change, particularly through policy experimentation.

Economic Transformation and Environmental Challenges Facing China

China's attempts to transform the economy date back to the late 1990s. At that time, top leaders announced their intention to shift the country's economic focus from extensive to intensive growth in the Ninth Five-Year Plan on National Economic and Social Development (State Council 1996). This plan sought to transform China's economic growth model from a dependence on the massive consumption of materials and resources to growth driven by the use of advanced technology and improved production efficiency. During decades of fast economic growth, it has increasingly depended on importing fossil fuels from other countries. China stopped being a net oil exporter in the early 1990s and became a fossil fuel net importer in the mid-2000s. It became a net importer of natural gas in 2007 and of coal in 2009; in 2014, it surpassed the United States to become the world's largest net importer of liquid fossil fuels (U.S. Energy Information Administration 2014; Wu 2014). Since 2016, it has been the world's largest producer of (and investor in) renewable energy (OECD 2020). Although China is a world leader in renewable energy investment, its annual coal consumption has not declined. In 2023, its newly installed solar PV capacity matched the total global addition in the last year, and its wind growth reached 66 percent year on year between 2022 and 2023 (International Energy Agency 2023). Yet China's total overall coal consumption has continued to increase during this period; its total amount of carbon emissions grew from 6,574 Mt CO_2 to 8,375 Mt CO_2 from 2010 to 2021 (International Energy Agency 2024).

China is home to one-fifth of the global population. Its rapidly growing urban population increased from 17.92 percent of the population in 1978 to 66.16 percent in 2023 (National Bureau of Statistics of China 2024), and the country has achieved unprecedented economic growth in recent decades. Yet severe environmental pollution and climate challenges threaten the sustainability of urban life. China experienced fast economic growth from its economic opening up in 1978: its annual gross domestic product (GDP) averaged over 9 percent from then to 2018; over the past four decades, nearly 800 million people have been lifted out of poverty (Ministry of Ecology and Environment 2019; World Bank 2024). This economic growth is attributed to rapid local industrialization, which has been fueled primarily by coal and other fossil fuels. Energy and industrial processes account for 48 percent of the country's total carbon emissions (International Energy Agency 2021). More than half of the 337 Chinese cities (prefectural level and above) failed to meet national air quality standards in 2019 (Ministry of Ecology and Environment 2019); this figure fell to nearly one-third by 2022.[1] Extreme weather and other climate change-related disasters have also taken a toll on the domestic economy and people's lives. According to the Ministry of Ecology and Environment, from 1984 to 2018, climate disasters caused a direct loss of 730 billion RMB (approximately 1.8 percent of annual national development growth, on average) and killed 120,000 people.[2]

Although subnational governments have been aware of the ecological tolls on the economy, environmental protection has not always been a high priority. In China's upwardly accountable political system, subnational governments have long prioritized economic growth over environmental protection, and periodic attempts to address environmental concerns have suffered from insufficient policy implementation and a lack of funds and capacity (Ding 2022, 160; Economy 2004, 91–92). After China opened up its economy and entered the World Trade Organization, an overemphasis on fast economic growth led to the abuse of natural resources and the environment (Li 2020).

In 2008, the National Development and Reform Commission created the Department of Climate Change; in 2010, the department introduced national low-carbon pilot city policy experimentation, which marked a key turning point in subnational climate engagement in China. A growing number of provincial-, municipal- and township-level governments announced their decarbonization goals along with specific action plans and targets during this time. From 2010 to 2017, there were eighty-one low-carbon pilot cities, six low-carbon pilot provinces, fifty-one low-carbon pilot industrial parks, more than four hundred low-carbon pilot communities, and eight low-carbon pilot townships (People's Republic of China 2023, 118–119).

Subnational government engagement in low-carbon policy experimentation soared in the late 2000s and early 2010s for three main reasons. First, top Chinese leaders began to pay more attention to climate and environmental issues. In 2009, President Hu Jintao first pledged to significantly reduce the country's carbon intensity compared to 2005 levels by 2020 (i.e., to consume fewer fossil fuels for the same amount of economic growth).[3] Premier Li Keqiang's declaration of a "war against pollution" in 2014 set the stage for strengthening environmental regulations. His 2014 report on the work of the government mentioned improving the capacity to address climate change coupled with reducing the country's total energy consumption as part of the measures for enhancing environmental and ecological quality (Li 2014). From 2016 to 2018, the national government increased its fiscal investment in environmental protection by 14.8 percent annually, for a total three-year investment of 2.45 trillion RMB including 47.4 billion RMB to address regional air pollution in major metropolitan areas (Liu 2019).

The second driving factor behind subnational climate engagement was that local government leaders (those at the provincial, prefectural/municipal, and township levels) became more responsive to climate and low-carbon issues at this time. In China's centralized political system, the political support of top local leaders (such as party secretaries and mayors) is essential. Local decision-making power is concentrated in the hands of those leaders (Eaton and Kostka 2014; Zhou 2008, 17). Research on officials' views of local government involvement in policy experiments suggests that the "insightful leader" is the main factor driving local areas and regions' policy experimentation (Yang 2011). Mayors of dozens of Chinese low-carbon pilot cities presented at the U.S.–China Climate-Smart/Low-Carbon summits in 2015 and 2016, pledging to lead their cities to attain their carbon peaks ahead of the 2030 national carbon peak pledge. Afterward, over eighty cities and townships, which account for 42 percent of the country's GDP, set carbon peak targets from 2017 to 2030 (Zhuang 2021, 74).

Third, institutional factors were also crucial in rising subnational engagement in decarbonization. The Department of Climate Change at the national government level was the first bureaucratic agency specializing in coordinating China's international climate negotiations and domestic climate policy development. It works closely with the National Center for Climate Change Strategy and International Cooperation, established in 2012, to promote national-level climate policies and engagement. These two organizations are the leading agencies strategically framing climate change as a local issue. They coordinated the implementation of the national low-carbon pilot city program in 2010, and have since helped build subnational governments' climate capacity by providing training for local officials and connecting local governments with international experts while they develop their low-carbon-city action plans. The national climate agencies have brought cities international recognition for their low-carbon practices at major international climate conventions, such as the United Nations climate change conferences.

Elevated national attention to climate and environmental issues can explain the increased subnational engagement, but not the divergent policy engagement outcomes. While all the pilot cities have developed low-carbon development action plans across the industrial, building, and transportation sectors, not all of them have created the policy incentives or project evaluation process needed to implement their low-carbon goals. Based on data from the Ministry of Ecology and Environment (2023b), of the eighty-one low-carbon pilot cities, 80 percent included carbon intensity in their annual government performance appraisals, but only 43 percent assigned specific carbon reduction targets to lower levels of government for implementation. While a small number of cities, such as Beijing, Shenzhen, and Shanghai, included hundreds of carbon-intensive entities in their regional carbon emissions trading systems (ETSs), over half of the pilot cities did not regularly account for local carbon emissions (Ministry of Ecology and Environment 2023b), which suggests huge local differences in climate engagement. How has subnational climate engagement been so diverse in such a centralized political system?

The next section explains the characteristics of subnational climate engagement in China and illustrates how subnational low-carbon policy experimentation has been crucial for climate policy implementation.

Characterizing Subnational Climate Policy Implementation in China

I study subnational climate policy implementation because no national climate pledges can be achieved without subnational engagement. Given that natural resource endowments and socioeconomic development vary significantly by region, it is often impossible to offer a one-size-fits-all approach to low-carbon policy implementation. Local governments have also been known for their role in impeding environmental policy implementation. Although China developed a comprehensive environmental law system in the late 1970s, local governments have long prioritized economic development over environmental protection, mainly due to the lack of public scrutiny and capacity, as well as frequent turnovers in political leaders (Ding 2022, 56–61; Eaton and Kostka 2014; Ran 2013). Climate policy implementation at the municipal level and below in China's fragmented bureaucratic system has featured initiatives being promoted without the necessary accompanying legislation, a lack of dedicated financial resources and staffing, and disruption caused by turnovers in top local government leaders. Research shows that local governments have had few incentives to implement climate policy initiatives beyond the planning stage (Qi et al. 2008; Song et al. 2021).

Policy experimentation in China is a unique, government-led process in which the national government grants considerable discretion to local governments to explore policy solutions to tackle rising policy challenges. In policy experiments, municipal and township governments can choose their policy instruments and set their own targets (Heilmann 2008b; Khanna et al. 2014); at the same time, they are expected to transfer local knowledge to the central governing body, and may receive political guidance, endorsement, and reaffirmation from the center.

Local government leaders' leverage in prioritizing the local policy agenda and mobilizing local resources has created two conditions for subnational climate policy experiments. First, local leaders can shape local responses to central political appeals. Although participating in policy experiments may not generate immediate political rewards, as the bureaucratic evaluation criteria would otherwise predict, local government leaders have been motivated to promote policy experimentation by the influence of political patronage, the need to address local challenges, and a desire for external recognition (Chung 2000; Wang and Yang 2021; Yang 2011). Top local leaders (e.g., mayors and party secretaries) wield supreme authority in shaping the priorities of the local policy agenda and making decisions related to economic and financial affairs (Eaton and Kostka 2014). The leading group set-up (*lingdao xiaozu*) is a useful coordination mechanism that facilitates local policy implementation by joining forces from key actors inside and outside the government (Chien and Hong 2018; Delman 2016). Local government leaders are aware of local social problems and are thus uniquely well positioned to command the necessary human and economic resources to address them (Yang 2011).

In a second condition, local governments have an incredible capacity to mobilize the resources needed to implement high-priority policies. For high-priority issue areas related to local economic development and industrialization, they can mobilize the necessary capital to foster local renewable energy firms' commercialization and mass production by providing tax breaks, land deals, and access to loans (Gallagher 2014, 29; Nahm 2021, 136–147). For instance, local governments used their authority to facilitate local solar PV enterprises' access to huge loans to help them quickly expand, and persuaded the national government to issue financial subsidies to help solar PV industries survive after international demand plummeted during the 2009 global financial crisis (Dong and Qi 2015). Local bureaucrats also work closely with key local actors to form alliances to facilitate policy implementation. Chen (2018, 16–17) finds that local bureaucrats' preferences and choices of alliances have led to different local approaches to development. The four case studies I analyze in this book illustrate that local bureaucrats play a crucial role in mobilizing the resources required for low-carbon policy experimentation.

The next section discusses how subnational governments have historically engaged to address energy, environmental, and climate challenges. It elucidates the process via which subnational engagement has transformed China's path toward decarbonization, from alleviating rural energy poverty and fostering renewable manufacturing competitiveness to laying the groundwork for China's climate governance.

Evolving Subnational Engagement in China's Low-Carbon Energy Transitions

Subnational low-carbon engagement since the late 1970s can be organized into distinct periods. From the 1970s through the 1990s, local areas and regions in China served as testing grounds for clean and renewable energy development with the goal of reducing energy poverty in remote and rural areas. In the 2000s, local governments served as champions of renewable energy manufacturing, facilitating the growth of domestic clean energy industries (Nahm 2017). During the late 2000s and early 2010s, addressing global climate change was translated as low-carbon development in local contexts. After the 2009 United Nations climate summit, when President Hu Jintao promised to incorporate climate change considerations into the country's economic and societal development plan, a growing number of local governments have expressed their interest in exploring low-carbon development, namely decoupling local economic growth from increased energy and resource consumption. They have sought to simultaneously tackle the climate issue and improve environmental quality during the economic growth pattern transformation. In the Xi Jinping era, China's low-carbon energy transition efforts have also benefited from its domestic battle with air pollution; climate issues have become synonymous with environmental protection since the climate institutional reform of 2018.[4] China's governance for addressing climate change has thus taken shape after almost a decade of subnational engagement in climate policy experimentation.

Initiating renewable projects designed to alleviate local poverty

The initial stage of local low-carbon actions coincided with the central government's emphasis on poverty alleviation and energy conservation from the 1970s to the 1980s. Early on, many new and renewable energy projects were focused on building small-scale hydro and wind power in rural and remote areas. These efforts were criticized for their energy inefficiency and ecological damage, but they helped local governments gain experience with clean energy development and demonstrated how it can help alleviate poverty.[5] By 2000, over 190,000 mini wind turbines (50 to 200 watts capacity) were in operation. Over 60 percent of these supplied power to people living in pastoral and coastal areas that previously had no access to electricity (State Council 2004, 44).

From the 1980s through the 1990s, local governments introduced renewable energy to a vast swathe of rural areas. The State Council established a Leading Group on Rural Energy to coordinate cross-divisional actions from several ministries involved in agriculture and hydro development, and they proposed using renewable energy sources to improve the electricity supply in rural areas. The council established four guiding principles: (1) the policy should be implemented according to local conditions, (2) multiple complementary energy sources should be harnessed, (3) those resources should be used comprehensively, and (4) efficiency should be pursued (*yindizhiyi, duonenghubu, zongheliyong, jiangqiuxiaoyi*) (Li 2019).

More than three hundred townships were engaged in the national-level "integrated rural energy development program" (*nongcun nengyuan zonghe jianshe xian gongcheng*), which introduced a variety of renewable energy options such as biomass, solar, geothermal, wind, and hydro power into local development. Leading groups were created to coordinate project implementation at the national, provincial, and township levels, and expert teams were established at the provincial level to supervise and evaluate the projects. These teams also offered guidance on project design, technological information, coaching on the use of the technology, policy research, and advice for key decision makers. As a result of this rural low-carbon energy development program, by 2000, over 8.5 million peasant households used biomass fuel, solar water heater applications reached 1,100 square meters, and the capacity of dam construction in rural areas reached 24.80 million kW with an annual energy production of around 80 billion kW (State Council 2004, 44).

Stunted development in subnational climate engagement

After introducing low-carbon projects in rural and remote regions, the central government began to support renewable industries in the 2000s to boost the domestic clean energy manufacturing capacity and evaluate feasible policy options for transforming economic growth patterns. The 2005 Renewable Energy Law codified the delegation of administrative power to local governments in renewable energy development. The law made local governments above the township level responsible for managing their jurisdiction's renewable energy development and utilization. Low-carbon initiatives are no longer restricted to remote areas as part of a central government effort to alleviate poverty; cities across the country are becoming increasingly engaged.

Municipal- and township-level governments have become instrumental to the fast expansion of local renewable energy business through their maneuvering of extensive administrative resources, such as permits and approval processes, land acquisition, loan guarantees, and preferential policies. Local governments in Jiangsu and Jiangxi provinces have worked closely with state-owned enterprises and banks to assist local solar PV enterprises in accessing capital to expand their manufacturing capabilities (Dong and Qi 2015).

China was slow to fully commit to reducing its domestic carbon emissions at the international level before the national government realized the political and economic benefits of the Clean Development Mechanism (CDM) (Lewis 2012, 16). Beijing was concerned that accepting international donations to fund domestic renewable energy projects might enable developed countries to meddle in its domestic greenhouse gas reduction obligations (Schröder 2012, 46). As local government engagement increased in setting up CDM centers for capacity building and project promotion, China became the home of the world's largest CDM projects in 2007 (Schröder 2012, 2; World Bank 2008, 26). Its wind power grew rapidly around this time. By the end of 2010, China had surpassed the United States in total installed wind capacity; twenty-nine out of thirty-two provinces and regions had developed wind farms, and Inner Mongolia accounted for 13.85 GW of the country's total 44.73 GW installed capacity (Li et al. 2012).

In the late 2000s, top leaders in Beijing began paying more attention to climate change—in line with the national strategy of scientific development, resource conservation, and environmental protection—which signaled political support for subnational climate engagement (National Development and Reform Commission 2008). With the launch of the World Wildlife Fund's "Low-carbon City Initiatives" project in 2007, cities such as Shanghai and Baoding began considering steps for decarbonization. These included improving energy efficiency in the building sector, local renewable industrial development, and international green technology transfer. More local governments joined forces and began to explore feasible low-carbon approaches inside and outside the government. From the beginning of 2008 to July 2010, over a dozen local governments demonstrated their interest in advancing low-carbon development (National Center for Climate Change Strategy and International Cooperation 2013, 3).

The formation of the National Leading Group for Climate Change, Energy Conservation, and Emission Reduction in 2007, and the creation of the Department of Climate Change in 2008 (and its affiliated think tank, the National Center for Climate Change Strategy and International Cooperation), created the basic institutional arrangements for local governments' engagement in climate issues. Eight provincial governments followed suit and established their own leading groups on climate change between 2007 and 2008 (Qi et al. 2008). National climate agencies worked with international agencies to help train specialized personnel in governments, universities, and research institutes, as well as enterprises. For example, the Deutsche Gesellschaft für Internationale Zusammenarbeit (German Corporation for International Cooperation) and the National Development and Reform Commission's Training Center introduced four local capacity-building training sessions in Hubei, Jiangxi, Jilin, and Shaanxi provinces from 2011 to 2014. The programs were designed to boost local governments' awareness of climate change and introduce the know-how and measures for local stakeholders to deal with carbon mitigation and climate adaptation. The three-day session held in Xiangyang City of Hubei Province trained over sixty officials from the development and reform commissions of six prefectural-level cities (National Development and Reform Commission and German Corporation for International Cooperation 2013). Because very few local experts specialized in climate change before then, local governments, in coordination with international experts, drafted most of the low-carbon pilot city work plans.

Beijing's increased interest in tackling climate challenges has triggered unprecedented local engagement in low-carbon initiatives. From 2010 to 2017, a growing number of cities created low-carbon pilot work plans; the majority sought to reduce carbon intensity ahead of the 2030 target. The number of national low-carbon pilot cities (and provinces) increased more than fivefold during this period from thirteen in 2010 to eighty-seven in 2017; as of 2017, almost every province had at least one low-carbon pilot city. Individual cities, such as Shijiazhuang, Nanchang, and Shenzhen, have passed low-carbon legislation (National Center for Climate Change Strategy and International Cooperation 2017, 1–36). In 2015, a couple of dozen local areas founded the Alliance of Peaking Pioneer Cities of China to exchange regional experience in low-carbon development. Its member provinces and cities committed to peaking their carbon emissions between 2018 and 2030.

Climate institutional reform and aftermath

In 2018, the national government announced its plan to transfer responsibility for climate issues (as well as the Department of Climate Change) from the National Development and Reform Commission to the Ministry of Ecology and Environment. This shift signaled that the central government was treating climate change as both an environmental and a development issue. For local policy implementation, the government needed to resolve the contradictions inherent in different management approaches that have been pursued in environmental protection and in tackling climate change. China's environmental protection policy enforcement has relied heavily on strengthening its environmental law systems and administrative measures from the top down, while low-carbon policy development has relied on fostering market-based mechanisms and policy experimentation from the bottom up (China Council for International Cooperation on Environment and Development 2019, 24–26).

To address these different regulatory and market management styles, the national climate institutional reform has been a deliberate political move to incorporate carbon emission reductions into the battle against air pollution to control the source of both greenhouse gas emissions and air pollutants. The central government stated that building a national ETS and related capacity building would still be a focus after the Department of Climate Change moved to the Ministry of Ecology and Environment, and that the Department would also take measures to support the development of low-carbon technology research and development, and to improve the carbon emissions standard systems for major industries.[6]

The bureaucratic reorganization related to climate issues also poses challenges in subnational climate policy implementation. Because few governments at the municipal level and below created climate change divisions before undertaking the climate institutional reform, they needed years to train specialized personnel for low-carbon policy implementation (China Council for International Cooperation on Environment and Development 2019). The COVID-19 pandemic has also strained local governments' ability to mobilize financial resources and investment (Qi et al. 2020).

In response to President Xi Jinping's carbon pledges in 2020, the central government announced high-level national guiding policies the following year, known as "1+N" policies, to mobilize decarbonization initiatives across multiple levels of government and policy sectors to achieve these goals.[7] The State Council requested that subnational governments issue carbon peak action plans with feasible timelines and implementation roadmaps (State Council 2021).

Subnational climate engagement will continue to shape China's approach to achieving its climate goals. In 2023, the National Development and Reform Commission announced a new plan to select one hundred cities and industrial parks as carbon peak and neutrality pilots to motivate local participation in exploring innovative policy mechanisms to achieve the national carbon peak and neutrality goals. The Commission requested that pilot cities and industrial parks detail their carbon targets across the energy, building, transportation, and forestry sectors, and asked provincial development and reform commissions and financial institutions to provide financial and policy support for the pilot projects (National Development and Reform Commission 2023b). Additonally, provinces, cities, and industrial parks have engaged in a variety of climate-related pilot programs to explore innovative ways to address policy challenges in fields such as climate finance, carbon emissions mentoring, climate adaptation, and synergizing air pollution reduction with carbon emissions reduction in the past decade (Ministry of Ecology and Environment 2022, 2023a). The extent to which China can achieve its climate goals largely depends on its climate governance capacity building, particularly local bureaucrats' ability to implement climate policies (China Council for International Cooperation on Environment and Development 2019).

Conclusion

This brief historical overview of subnational engagement in China's low-carbon energy transitions reveals that local areas and regions have long been at the forefront of experimenting with low-carbon ideas and solutions linked to poverty alleviation, the development of renewable industries, and addressing climate change. Climate change issues did not catch local governments' attention until the late 2000s, when President Hu Jintao first pledged to the international community to contain domestic carbon emissions. After decades of rapid growth at the expense of the environment, the country's national development strategies sought to better align economic expansion with environmental protection. The development of low-carbon initiatives over the last decade can largely be attributed to the elevated national political attention to environmental protection and the strategic incorporation of decarbonization into the country's domestic economic transformation.

While subnational governments face political and financial resource constraints, they have taken a leading role in developing innovative solutions to balancing the competing goals of environmental protection and economic development during the decarbonization process. For example, the low-carbon initiatives not only deal with local economic challenges, such as energy poverty alleviation, but they also shape national responses to addressing climate change, such as CDM projects and renewable industry development.

Subnational engagement in climate governance began after climate issues emerged as a distinct policy area in the late 2000s, starting with low-carbon policy experimentation. For the national low-carbon pilot city program initiated in 2010, no formal policy or institutionalized financial subsidy schemes were issued; municipal governments were given considerable leeway to design their own workplans to build low-carbon cities across the industrial, transportation, and building sectors. Local climate experimentation and implementation have faced challenges caused by frequent reshufflings of national-level authority over climate issues and local economic stress during and after the COVID-19 pandemic. Nevertheless, subnational climate engagement persists in terms of creating low-carbon action plans, experimenting with low-carbon regulations, and carbon emissions measurements. Despite the progress in making action plans, many barriers to improved local implementation persist, including a lack of dedicated staff, institutions, and financial resources on climate issues (particularly in local bureaucracies), as well as an unclear governing structure.

Subnational governments are on the frontline of implementing China's decarbonization. Local low-carbon policy experimentation is thus a fascinating and important place to study the country's evolving approach to addressing climate change. Climate policy experimentation continues as the deadlines for the carbon pledges approach. More subnational governments have engaged in climate experimentation, and their participation will continue to diversify both their own engagement outcomes over time as well as subnational climate engagement as a whole.

The next chapter elucidates the mechanism behind uneven subnational engagement in low-carbon policy experimentation and the conditions under which local governments—through such experimentation—can foster policy institutionalization toward decarbonization.

3

A Leadership Approach to Low-Carbon Policy Implementation

China's launch of the national low-carbon pilot city program in 2010 was a turning point that inspired a wide range of subnational governments, in both developed and developing regions, to explore growth paths that were less dependent on the increased use of fossil fuels. Based on data from the Ministry of Ecology and Environment (Ministry of Ecology and Environment 2023b), over a third of the eighty-one low-carbon pilot cities have issued low-carbon action plans since then, yet only 5 percent have pushed legislation on low-carbon-related issues through local people's congresses. Sponsoring low-carbon policy experiments requires mobilizing political, economic, and human resources inside and outside the government. It can sour relations with carbon-intensive industrial sectors and incur opposition, making it a demanding policy choice for local governments. Why have cities embarked on low-carbon policy experimentation? Translating low-carbon pledges into actions requires the continued mobilization of implementing resources, but changes in local political leaders can disrupt this process. How have some cities sustained stronger policy engagement than others despite political turnovers?

I argue that bridge leaders in the local government drive the process of securing the political support and implementing resources required to institutionalize low-carbon policy. I contend that local governments are more likely to initiate and maintain low-carbon policy experiments when bridge leaders—who are often mid-level bureaucrats in local governments—can secure political support and forge implementation alliances to introduce institutional efforts toward decarbonization throughout the implementation process.

This argument departs from most existing theoretical explanations by highlighting the importance of bridge leaders' ability to *mobilize resources* in the implementation process rather than the bureaucratic structures and incentives in which the leaders are embedded. Due to the opacity of China's government policymaking to outsiders, local implementation processes can

Implementing a Low-Carbon Future. Weila Gong, Oxford University Press. © Oxford University Press (2025).
DOI: 10.1093/9780197757451.003.0003

be difficult to investigate. I overcome this challenge to answer the questions raised in this chapter using comparative and qualitative methods, consisting of comparative case analysis, process tracing, analysis of original government documents, and expert interviews, that allow me to reconstruct the policy process of low-carbon experimentation from agenda setting to implementation. I illustrate the local decision-making process and how these leaders win political support and build implementation alliances, which are key to local engagement in policy experimentation. I triangulate the interview data with documentary sources including government policy documents, program reports, statistical yearbooks, and media coverage. The theoretical framework outlined here is a useful tool to help interpret the empirical evidence presented in the case chapters.

Limitations of Existing Explanations

Scholars of climate and environmental politics and policy have long sought to explain subnational policy engagement in addressing climate and environmental challenges (Bulkeley 2013; Vogel 2018; Rabe 2004; Hughes 2019). Previous studies of central–local relations and environmental politics in China generally emphasize that subnational government responses to national policies are influenced by bureaucratic incentives from the central government (Bulman 2016; Li and Shapiro 2020; Oi 1999; Zhou 2008). This section briefly reviews two major theoretical approaches to assessing subnational policy actions—the principal–agent model and the bureaucratic structure model—before I introduce my own.

The *principal–agent model* explains local policy responses to central political appeals. Researchers who employ this model often assume central–local relations in China are a zero-sum game and portray local governments as rational and self-interested actors in a centralized political system. According to this approach, local officials only respond to central commands via local initiatives because they perceive the benefits of doing so (promotion or personal financial gain) as being guaranteed; otherwise they may either implement central requests halfheartedly or advance their own interests at the expense of those at the center (Bulman 2016; Chung 2000; Oi 1999). The model is less effective at explaining how (or why) local officials might respond to low-carbon policy experiments, especially if they are neither institutionalized within the policy system nor have measurable criteria linked to bureaucratic evaluations. Early work on the motives of local engagement in policy experiments suggests that local governments are unlikely to engage in experimental programs for the supply of public goods because there is less chance of financial gains (Heilmann 2008b). Recent work on policy innovations echoes the emphasis on the central–local power struggle but suggests it is important to understand how local strategies of engaging in policy experiments affect good governance and social welfare (Teets and Noesselt 2020).

The *bureaucratic structure model* emphasizes the distribution of authority and resources within the fragmented bureaucratic system. It highlights that local governments may have different responses to central appeals depending on the distribution of authority and financial resources within the bureaucratic system for implementing certain policy issues (Lieberthal and Oksenberg 1988). In contrast to the principal–agent model, the bureaucratic structure model maintains that local responses to central directives are *not* solely determined by officials' rational calculations for maximizing beneficial returns; how local leaders assess the local conditions and how they believe the central policy will affect local interests also play a role (Lampton 1992). Taking the bureaucratic structure into account helps us understand that local policy responses are not merely the result of local leaders' rational calculations (Xue and Zhao 2018); they are also the product of their responses to competing priorities, assigned at the central level, given the constraints of local political and economic resources (Lampton 1992). But it is still hard to understand why subnational governments with seemingly similar bureaucratic structures may generate such different responses in low-carbon policy experiments.

Both models posit that local engagement in policy experimentation is primarily determined by the incentive structure provided by the central government. While both recognize local leaders' capacity to assess and respond to structural opportunities, neither approach sufficiently considers the structure of local government leadership as a necessary element for understanding local decision making. Nor do they investigate how local leaders merge ideas and actions in the policy process. Principal–agent analysts often pay little attention to the local policy process or the evolution of institutional structures in response to political uncertainty. While focused more explicitly on the policy process, bureaucratic structure analysts do not systematically examine interactions between local leaders and the institutional structures in which they work.

The primary difference between the two dominant models is that principal–agent analysts argue that local responses are the result of local officials' calculations of their individual political and economic returns, which are often shaped by national government incentives. Bureaucratic structure analysts instead emphasize that actions are influenced by the distribution of authority and resources, determined by the national government, through the bureaucratic structures in which local officials operate. Another noteworthy difference is that principal–agent analysts tend to ignore the bureaucratic struggles over resources and authority as a key feature of local policy processes, while bureaucratic structure analysts do not thoroughly examine the resource mobilization process, which local leaders often facilitate to eliminate constraints and expedite policy implementation (Heilmann and Perry 2011, 22).

Taking central–local relations and bureaucratic structure into account helps us explain why local governments respond to central appeals in a centralized political system like China's. But it is still unclear why local governments participating in low-carbon policy experimentation at roughly the same time have developed divergent paths toward policy implementation. The next section introduces another approach to unravel this puzzle.

The Leadership Approach

My conceptualization of leadership differs from conventional approaches focusing on implementing policy from the top down in that it considers policy experimentation a local resource mobilization process that translates policy ideas into actions. This means knowing how local bureaucrats maneuver the bureaucratic structure to mobilize the necessary implementing resources for low-carbon policy experiments can be as important as understanding the influence of central–local relations and the bureaucratic incentives that are often largely set by the national government.

This approach builds on previous studies that highlight the importance of leadership, resource mobilization, and alliances in the policy process (Hughes 2019, 69–71; Morris and Staggenborg 2007; Perry 2011; Rabe 2004, 22). Prior research points out that local leaders and the resource mobilization process are central to understanding local decision-making processes (Bulman 2016; Chung 2000). This book goes beyond previous work to explore the leadership mechanism behind resource mobilization in the process of subnational policy implementation. Unpacking the resource mobilization process is especially important for advancing scholarly understanding of the role of local government leadership in the implementation process beyond the limits of principal–agent or bureaucratic structure analysis.

The leadership approach shifts the scholarly focus (1) from central–local relations to the local decision-making process and (2) from political leaders' role in setting the local agenda to that of mid-level bureaucrats who serve as implementation bridges in initiating and sustaining low-carbon policy experiments. This study parallels other scholarly works that emphasize the importance of local bureaucrats in promoting subnational initiatives in China's centralized political system (Ang 2016; Chen 2018; Heilmann and Perry 2011; Schröder 2012). But it goes further to point out the bridging role of mid-level bureaucrats, which is often overlooked in local policy decision-making processes, especially those related to introducing policy changes and influencing implementation outcomes.

I argue that local bridge leaders influence the implementation of low-carbon policy experimentation in three important ways. First, mid-level bureaucrats who occupy leadership positions in powerful agencies can mobilize the necessary political support to inspire action on decarbonization within the government. Second, bridge leaders configure implementation strategies by building alliances with key actors inside and outside the government to leverage human and non-human resources and provide policy solutions via experimentation. Third, such leaders work with their allies to build new low-carbon institutions out of the policy experimentation process, which helps maintain policy experiments during local political turnovers. Low-carbon policy institutionalization built on formal and less formal city-level climate institutions also gives stakeholders venues for providing new incentives and ideas for continued low-carbon policy engagement (Hughes 2019, 69). Bridge leaders' strategic efforts make continued local engagement in low-carbon policy experimentation more likely, particularly since the creation of formal institutions (standards, laws, regulations, implementing agencies) makes policy implementation less vulnerable to changes of individual leaders.

The Structure of Bridging Leadership

Acknowledging the conceptual complexity and debate about "great man theories of history" (Li and O'Brien 2008, 1), I define *Bridging Leadership* as the process by which local bureaucratic entrepreneurs broker the political support and implementation alliances required to translate low-carbon ideas into policy actions (see Figure 1.1). This definition departs from both conventional perspectives on political leadership that focus on the characteristics of individual leaders, and from more mainstream conceptualizations of climate leadership in the study of climate politics and policies, which focus on the leader–follower relationship between countries in climate policy diffusion or the leading role of agenda setters in shaping the direction of international climate negotiations (Schreurs and Tiberghien 2010; Steinbacher and Pahle

2015; Torney 2015). I follow those who emphasize that the study of leadership should go beyond the structure–agency dichotomy that focuses on either the role of the individual leader or on how external constraints affect the emergence of leadership (Morris and Staggenborg 2007). By examining how local government leaders mobilize resources, this study considers how local governments create favorable conditions for introducing institutional efforts toward decarbonization in the policy process.

Political leaders

Political leaders—often the party secretary or mayor in Chinese cities—hold supreme political authority within the local government and are responsible for local policy decision-making. While democratic political systems do not restrict where a policy advocate may emerge (Kingdon 1995, 179–184), in a centralized political system like China's, access to the policymaking process is limited to a closed group of people—namely political leaders and those who work closely with them (Lampton 1992). These political leaders are important for local agenda setting even though they generally serve in a jurisdiction for no more than five years before being rotated to a new post (Eaton and Kostka 2014).

Effective local low-carbon initiatives require political support to ensure they are high on the government's policy agenda (Bulkeley et al. 2015, 10; Jiang 2015; Yang 2011). Yet the turnover of political leaders when they are assigned to a new posting elsewhere can thus easily cause an issue to fall off the agenda (Eaton and Kostka 2014). Chinese local officials' motivation to engage in policy innovations can range from personal idealism to career advancement (Fewsmith 2013, 155). In Simon's (1983) study of the human behavioral mechanisms behind decision making in bounded rationality, he argues that people focus on certain problems at particular times. Whether or not a person would prescribe particular solutions to a problem depends on (1) their experiential familiarity with the issue, which informs the range of options, and (2) the emotions aroused that lead to action. According to this approach, a local government's policy engagement on a new issue like climate change can be influenced by its leaders' exposure to the issue area and their willingness to pursue low-carbon solutions to address local development challenges.

Given the highly technical and scientific complexities associated with certain policy issues, political leaders need experts and their networks to succeed (Helco 1978, 118). The availability of expertise may enhance or diminish political leaders' ability to introduce new policy issues or to reframe existing policy priorities. Bureaucrats, together with field experts acting as policy advocates, often specialize in preparing proposals and finding solutions for policy problems; government leaders tend to be generalists and thus rely on this expertise for policy implementation (Kingdon 1995, 205).

Bridge leaders

Bridge leaders make decisions in their specialized policy fields and have the authority to command implementing resources in the bureaucratic agencies they oversee. Mid-level local bureaucrats orchestrate the implementation of low-carbon policy experiments. The extent to which these bridge leaders are fully engaged in mobilizing a policy action can impact the local leadership's capacity to build alliances with other entities (Morris and Staggenborg 2007). Thus, the more able bridge leaders are to broker political support from top political leaders and implementing resources from key local actors, the stronger local engagement in policy experimentation is likely to be. Bridge leaders' ability to secure political support from top leaders and mobilize implementing resources directly influences whether local governments can translate their policy ideas into feasible actions.

Given their strategic position, bridge leaders are well positioned to create favorable conditions for local policy initiatives by building local implementation alliances or accessing resources from higher levels of government. Two skills are essential: (1) using rhetoric to frame an issue to gain political and public support and (2) building alliances with key actors and groups who might have interests at stake in the proposed action (Doig and Hargrove 1990, 14–15). These local officials and specialists come from diverse personal and professional backgrounds, which serves as a counterbalance to the biases of political leaders, who lack the issue-specific knowledge needed for decision making. By offering policy consultancy, bridge leaders with policy insights and specialized local knowledge are a credible and reliable source of information for decision making. Rabe (2004, 23–24) theorizes that these specialists usually command a strong network related to an issue area, and thus function as key brokers for coalition building as well as content experts. This brokering function, and its role in forming implementation alliances, helps mobilize resources in support of local policy implementation.

In summary, two key components are central to the structure of bridging leadership: the top political leader and the bridge leader. Interactions between a locality's political leader and bridge leader are crucial to developing a consensus on strategies to link the numerous political issues with policy solutions. Where government leadership holds regular meetings and discussions as part of the implementation process, policy experiments may proceed rapidly (Tie et al. 2020). Yet if internal leadership interactions between political leaders and bridge leaders are weak, or if there is a change in the political leader, the policy experiments may proceed much more slowly. While prior studies have demonstrated that political turnovers have been the key obstacle to sustaining policy experimentation in China (Eaton and Kostka 2014; Fewsmith 2013, 84), my research demonstrates that low-carbon policy initiatives can survive changes in government leadership if a sufficient level of policy institutionalization is created during the experimentation process.

Resource Mobilization

Local governments' resource mobilization capacity shapes the outcome of policy initiatives because it determines the extent to which local governments can translate what was promised or proposed into real actions. I define *resource mobilization capacity* as a local government's ability to leverage or create institutional structures to channel, recruit, and consolidate implementing resources (including collaboration among key policy actors) to enact policies.

Local government leaders' ability to engage key local actors in the policy process is particularly useful for resource mobilization. The strength of the leadership's resource mobilization capacity is influenced by the richness of resources available to it, as well as the structural factors that shape leaders' ability to access the most crucial resources for problem solving (Ganz 2000). The more institutional structures bridge leaders build to mobilize implementing resources (such as trained personnel and implementation alliances), the stronger local government policy engagement is likely to be. The rest of the section describes the two key components of resource mobilization: institutional structures and implementing resources.

Institutional structures

Institutional structures are created when the local leadership builds institutional connections with key actors to recruit implementing resources and build alliances for policy implementation. I follow Ikenberry's (1988, 192) description of how institutional structure affects government actions: "The state is not simply an area of political conflict; it is an institutional structure that gives government officials a unique standing in the pursuit of their own goals and strategic perspectives." The function of these institutional structures is relevant to government intervention in the policy experiments studied here as well. Such structures serve as a bridge between actors through formal and informal connections—such as leading groups and government networks—that enable leaders to coordinate cross-division cooperation within the government, as well as collaboration with social, business, and industrial actors for policy implementation (Gilley 2017; Nahm 2017).

Institutional structures connect relevant actors with the goal of completing specific organizational tasks and channeling the necessary resources (Manion 2015). The configuration of institutional structures can determine which actors or agencies are involved in the policy process, and therefore the distribution of the authority and resources that are under local leaders' control. Gilley (2017) argues that executive administration is more likely to be collaborative at the local than the central level due to its relatively small size; thus, local governments are a good place to craft new institutional structures to mobilize cross-divisional efforts for decarbonization.

Institutional structures can also include government-affiliated agencies and non-governmental organizations. The local government exercises primary control over the process of building networked connections outside of the government. Having more bureaucratic connections does not necessarily lead to better policy performance if there are parallel connections of competing interests (Mai and Francesch-Huidobro 2015, 254–268). Indeed, conflicts of interest between networks can hinder policy implementation, but leading groups can often mediate such conflicts among bureaucratic agencies and coordinate work across them in support of the policy process. Building institutional structures is important because they enable local government leaders to mobilize and redistribute resources inside and outside the government in support of policy actions. To better understand Chinese cities' low-carbon policy processes, it is thus important to consider the extent to which government leaders can establish institutional structures to coordinate actors and mobilize resources in support of policy implementation.

Implementing resources

This study builds on past research, which has broadly framed resources as "a reserve source of supply or support" (Edwards and McCarthy 2007), primarily in the context of mobilizing collective action within democratic governments. In China, bureaucratic officials are the main actors in government-led mobilizations (Perry 2011, 52). I focus on two types of resources that are critical to advancing low-carbon policy initiatives: trained personnel and local implementation alliances. I refer to these as *implementing resources* as they help ground the proposed policy initiatives in tangible practices to move the policy process forward.

Trained personnel play an essential role in helping the leadership team draft policy directives, organize government meetings, implement the proposed policy solutions, and supervise implementation outcomes. There are two sources of trained personnel: (1) government officials and (2) non-governmental agencies such as businesses, universities, and research and financial institutions.

Chinese bureaucrats generally specialize in administration, finance, or strategic planning. Those specializing in strategic planning can be trained in low-carbon issues if they are affiliated with climate change divisions or local bureaucrats in charge of low-carbon issues. Appointing trained personnel to the executive office allows outgoing action directives and incoming action feedback to be reviewed by officials with specialist knowledge. They may even act as bureaucratic entrepreneurs who can influence the decision-making process.

Other trained personnel work in non-government units that connect the government and business sectors in specific policy areas. Their low-carbon knowledge stems from either their professional expertise or prior involvement in low-carbon activities. Non-government specialists are sometimes seconded to a government administrative unit to assist with projects in their area of expertise. In two of the four case studies—Shenzhen and Xiamen—loaned personnel from local research institutes worked on low-carbon policy issues in the local government for one to two years. Such ties between government and non-government agencies also indicate how local governments mobilize external human resources in policy implementation.

In the case study chapters, I will show that leaders rely on trained personnel to create detailed and feasible work plans. The extent to which employees understand the content and contours of the proposed policy goals influences their ability to develop solutions tailored to local conditions. While not every local area and region has the capacity to form a specialized agency from the start, as the policy develops it requires local governments to build a group of professional staff with expertise in a specific policy domain to implement policies that require an advanced level of coordination inside and outside the government (Ang 2016, 40-42). Enhanced connections among trained personnel can facilitate the creation of a division of labor and help scale up actions in the long run. As Hoffmann (2011, 158) noted, "A division of labor process or outcome can and must be fostered in experimentational governance system."

The second type of implementing resource is *implementation alliances*. The leadership team's success hinges not only on its own expertise, but also on how effectively it interacts with (and influences) other key individuals and organizations to take action (Tucker 1995). In policy process research, building alliances is seen as particularly important for emerging issue areas because it causes policy advocates to incrementally promote an issue through those who support and foster the policy, and to gradually boost the political significance and support of key actors without challenging vested interests in the political system (Baumgartner and Jones 2009, 100–102). Therefore, building implementation alliances is important for leaders who seek to introduce changes into the existing policy system.

Local leaders who work on promoting a new issue area may not necessarily need to diffuse the opposition, as the vested interests have yet to see the new issue area as a challenger. There may be a greater need to engage relevant bureaucrats and non-government units who might be apathetic to the issue for allied actions. Bureaucrats with complementary field expertise provide information about the possibility of policy integration and even an exchange of resources, which allows the leadership to target the appropriate issue area to achieve progress. This co-opting mechanism is a useful way for local governments to organize multi-agency coordination to orchestrate local low-carbon policy implementation (Gilley 2012).

Measuring Local Engagement in Policy Experimentation

Although local governments have traditionally been considered the key barriers to environmental policy implementation, they have emerged as leading actors in low-carbon experimentation and will ultimately determine the success of China's low-carbon energy transitions in two ways. First, the central government does not stipulate specific policy goals for low-carbon pilot cities; local governments propose their own objectives and create detailed action plans to experiment with policy solutions that the national government has not previously tested. For example, local governments have led on monitoring carbon emissions and experimenting with new policy instruments, including data management platforms and accounting systems (Xu et al. 2017, 3). However, local governments can choose to cancel policy experiments without notifying the central government if the necessary political support is no longer available (Heilmann 2008b).

Second, local low-carbon initiatives have developed where the higher level of government has provided neither institutionalized financial subsidies nor guaranteed political and economic returns. The central government encouraged local governments to engage in low-carbon policy experiments without providing additional funding (Lo et al. 2019; Song et al. 2020; Tie et al. 2020). Thus, local governments have had to reallocate their own political and economic resources as the policy experiments unfolded. Indeed, nearly thirty pilot cities and provinces have established dedicated funds for low-carbon development and emission reductions, for instance, to finance technological innovation as well as low-carbon pilot projects (Xu et al. 2017). Local government leaders have played a crucial role in creating government coordinating mechanisms and specifying detailed targets and task forces for local bureaucratic agencies.

Prior studies on policy experimentation in China have acknowledged that policy experiments have the potential to trigger major policy change (Teets and Hurst 2015, 1; Teets and Noesselt 2020), but have only vaguely measured local policy engagement. As Fewsmith (2013, 39) pointed out, "Institutionalization, however, differs from institutional innovation. There is a lot of institutional innovation in China, but very little institutionalization of those efforts." Many scholars have attempted to understand low-carbon pilot city programs' performance through index evaluations (Khanna et al. 2014; Pan et al. 2013; Shi et al. 2018; Yang et al. 2018; Yu et al. 2019). Yet prior studies have paid less attention to the policy processes local governments use to engage in institutional innovations and institutionalization related to decarbonization. The methodological challenges associated with understanding the performance of local low-carbon policy experimentation include a lengthy, iterative policy process (Pressman and Wildaysky 1984; Sabel and Zeitlin 2012), the lack of transparency in China's energy accounting methods (Lewis et al. 2015), opaque formal evaluative criteria used by the central and local governments (Teets and Hurst 2015), and the difficulties of quantifying the amount of local effort put into policy experiments (Hoffmann 2011; Huang and Zhou 2023). To address these challenges, I use a process-oriented, rather than a results-oriented, approach to evaluate local policy performance in terms of local *policy engagement*. To evaluate the variation in local policy engagement in translating low-carbon promises into actions, I consider the level of low-carbon policy institutionalization, which reflects the transformed incentives and constraints involved in local policy decision making.

Following North (1990), I distinguish between formal and informal institutions when considering institutional change. *Formal institutions* are systems devised to shape human interactions, such as legislation and regulations.

Because they are fairly stable, changes in such institutions can have a major effect on policy experiments. When policy instruments tested in policy experiments become embedded in formal institutions—such as a locality's five-year plan or local environmental legislation—they can facilitate long-term policy changes by legitimizing the local leadership's proposed goals and actions, as well as diffusing potential opposition. *Informal institutions* are "codes of behavior" inherent to a specific political, economic, or cultural environment (North 1990). They shape the relationships and behavior patterns of those involved in the policy process. Informal institutions in China have been found to play an important role in mobilizing the necessary political and economic capital to increase the government's adaptive capacity (Tsai 2004). The introduction of "marginal institutional changes and informal institutional arrangement" into the policy experimentation process often reflects how power struggles have been mitigated in local policy implementation (Fewsmith 2013, 39).

Informal institutions have been used to boost China's environmental protection as a campaign style of policy implementation that focuses on the short-term policy instruments used to introduce incremental improvements in local environmental policy enforcement (van Rooij 2006). Such institutions can present unexpected channels within the policy process to distribute authority and resources. New experimental programs that operate under this type of informal institution often serve as the starting point for policy experiments, which may later lead to formal institutional changes in the policy process (Heilmann 2008a). The policy instruments that these experiments promote may become integrated into the policy system as implementation alliances expand (Jänicke 2012, 27). Thus, local engagement in low-carbon policy experimentation can be evaluated using two indicators: (1) the level of low-carbon policy institutionalization (i.e., the extent to which low-carbon issues have been integrated into the existing policy system via regulations, standards, and legislation together with the creation of specialized implementation agencies with full-time staffs) (Jänicke and Jacob 2004) and (2) the continuity of experimental programs.

Conclusion

While local governments are constrained by finite political and economic resources, they can create the political support and implementation alliances needed for low-carbon policy experimentation. Bridge leaders mobilize political support from political leaders and broker implementing resources from key actors inside and outside the government. The level of local engagement in low-carbon policy experiments is largely influenced by the local bridging leadership's ability to win political support and mobilize the necessary implementing resources to institutionalize the efforts toward decarbonization.

For scholars of central–local relations, the leadership approach I have developed suggests that local mid-level bureaucrats play a bridging role in low-carbon policy implementation. I argue that the availability of local bridging leadership and their mobilization capacity can be more important than bureaucratic incentives in initiating and maintaining low-carbon policy experiments, and that the level of local government engagement in such experiments is better understood through the resource mobilization process than local leaders' rational calculations.

For bureaucratic structure theorists, my approach helps explain the conditional effect of bureaucratic structures on local government leadership. This book's findings and theory suggest that such structures influence the distribution of resources and authority that might be accessible to local governments for low-carbon policy experimentation. But the structure is not static: some governments can extend their institutional connections with key actors that draw on outside political, financial, and legislative resources through the implementation alliances they mobilize to institutionalize low-carbon policy experiments. In other words, bureaucratic structures are important, but bridging leadership in local government is *real.* By focusing on bridging leadership, I argue that future studies can better appreciate the resource mobilization process that local governments can use to maneuver the bureaucratic system, overcome the constraints of political and financial resources, and propel the low-carbon policy implementation process.

4

Substantive Engagement

Shenzhen

Shenzhen is a trailblazer for low-carbon policy experimentation in China. It has demonstrated substantive engagement, backed by strong political support and implementation alliances, to translate low-carbon ideas into sustained policy actions. It was the first low-carbon pilot city to launch a local carbon emissions trading system (ETS). Over the years, Shenzhen's municipal government has built a flurry of low-carbon institutions including standards, regulations, and legislation, and created a division of climate change in the government. The municipal government leadership led exemplary policy programs such as the Shenzhen ETS and the Shenzhen International Low-Carbon City Program in the experimentation process. The bridge leader—a vice mayor who spent most of his career in Shenzhen—played a central role in mobilizing political support and implementing resources by building alliances with key actors inside and outside the government. The bridging leadership allowed Shenzhen to translate proposed policy pledges into actions, resulting in the creation of new low-carbon institutions. In this way, the city sustained its low-carbon policy engagement even after changes in local government leadership. This chapter traces the implementation of Shenzhen's low-carbon pilot city initiatives from 2010 to 2018.

Transforming Local Economic Growth through Low-Carbon Development

Shenzhen was China's first special economic zone. It is located at the mouth of the Pearl River in Guangdong Province, close to Hong Kong. It is home to over seventeen million people (Statistics Bureau of Shenzhen Municipality and Survey Office of the National Bureau of Statistics in Shenzhen 2024). Since 2013, 80 percent of the city's energy needs have been supplied by nuclear, natural gas, and electricity (Development and Reform Commission of Shenzhen Municipality 2012). From 2011 to 2019, its energy consumption

Implementing a Low-Carbon Future. Weila Gong, Oxford University Press. © Oxford University Press (2025).
DOI: 10.1093/9780197757451.003.0004

per unit of GDP decreased by 3.2 percent to 5.1 percent each year (Statistics Bureau of Shenzhen Municipality and Survey Office of the National Bureau of Statistics in Shenzhen 2019, 113).

Shenzhen's engagement in decarbonization is rooted in its efforts to restructure local economic growth. In the late 1990s, local authorities recognized the environmental damage caused by rapid industrial expansion, but the municipal government did not explicitly express a willingness to tackle this problem until the 2000s. In 2005, the Party secretary of Shenzhen municipality, Zongheng Xu, identified three threats to local sustainable growth: the scarcity of land, water, and energy; population growth; and the lack of energy and environmental resources. Improving energy efficiency and curbing the consumption of local resources then became priorities in efforts to transform local growth patterns (Chen et al. 2013). Shenzhen government efforts to transition away from dependency on labor-intensive and resource- and energy-heavy industries redoubled in response to the global economic crisis of 2009 (Xu 2009). The government tightened local environmental regulations, causing labor-, energy-, and resource-intensive enterprises to move out of the city and the Pearl River Delta area.[1] Local leaders' attention to environmental challenges highlighted the political significance of these issues and energized local interest in the environment, which made it possible to embrace rapid change and justify immediate policy actions.

Shenzhen was particularly active in low-carbon policy activities after it became a national low-carbon pilot city. Whereas more than eighty pilot cities have developed action plans and set targets, Shenzhen is one of the few that have initiated and maintained local low-carbon policy engagement by working with a variety of key actors inside and outside the government, as well as across different levels of government. It established a leading group in 2010 to address climate change and coordinate low-carbon policy implementation. Chaired by Mayor Qing Xu, the leading group involved more than twenty bureaucratic departments and working units. Shenzhen's Development and Reform Commission laid out its own Low-Carbon Development Medium- and Long-Term Plan (2011–2020), which focused on building a low-carbon industrial structure; a dedicated low-carbon development fund allocated more than 50 million RMB (US$8 million) each year to technology innovation, low-carbon businesses, and research and professional recruitment during this period (Shenzhen Municipal People's Government 2012).

While its total energy consumption has not declined, the Shenzhen government achieved its goal of increasing the contribution of low-carbon industries to local economic growth. The city's energy intensity level was 0.35 tonnes coal equivalent/per 10,000 RMB in 2018, putting it ahead of its 2020 target (Statistics Bureau of Shenzhen Municipality and Survey Office of the National Bureau of Statistics in Shenzhen 2019; see Table 4.1 and Figure 4.1). The share of heavy industry in the local economy decreased by 25 percent between 2008 and 2016, while the share of low-carbon tertiary industries increased by 41 percent during this period (Statistics Bureau of Shenzhen Municipality and Survey Office of the State Statistics Bureau in Shenzhen 2017, 10).

Shenzhen's pioneering role in low-carbon policy experimentation has largely been driven by its bridging leadership. Vice Mayor Jie Tang, acting as the bridging leader, worked closely with the mayor to bring together heads of government bureaucratic units and experts from outside the government, and to obtain the support of the Shenzhen Municipal People's Congress to promote major low-carbon policy experimental programs and support their implementation by creating low-carbon institutions such as legislation, regulations, and standards as well as a group of trained personnel for enforcement.

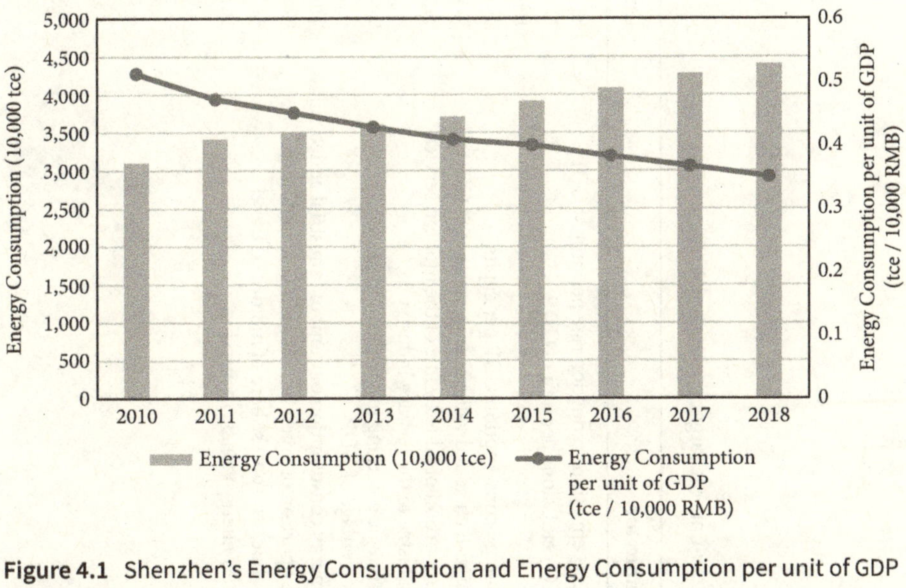

Figure 4.1 Shenzhen's Energy Consumption and Energy Consumption per unit of GDP (2010–2018)

Note: Shenzhen's energy consumption per unit of GDP for 2010–2013 is cited from the Shenzhen Statistical Yearbook 2019 and data for 2014–2018 are calculated by the author using figures from the same source. Shenzhen's total energy consumption for 2012–2018 is also cited from the Shenzhen Statistical Yearbook 2019, and the data for 2010 and 2011 are estimated by the author based on annual changes in energy consumption reported in the same source.
Source: Statistics Bureau of Shenzhen Municipality and Survey Office of the National Bureau of Statistics in Shenzhen (2019).

Table 4.1 Shenzhen's Low-Carbon Targets: 2015 and 2020

Item	No.	Indicator	Unit	2015	2020
Low-Carbon Production Output	1	CO_2 emissions reduction rate per unit of GDP	%	39	> 45
	2	CO_2 emissions per unit of GDP	Tonnes of carbon dioxide/10,000 RMB	0.90	0.81
	3	Energy consumption GDP per capita	Tonne coal equivalent/10,000 RMB	0.398	0.366
	4	High-tech added value as a percentage of GDP	%	35	40
	5	Proportion of added value of the modern service industry to the tertiary industry	%	60	65
	6	Strategic emerging industries' added value as a percentage of GDP	%	40	50
	7	Energy consumption per unit of industrial added value	Tonne coal equivalent/10,000 RMB	0.394	0.355
	8	Proportion of green buildings in new buildings	%	40	80
	9	Public transport's share of motorized travel	%	56	65
	10	New energy vehicle ownership	10,000 cars	5	10

Low-Carbon Resources	11	Proportion of non-fossil energy in primary energy consumption	%	15	> 15
	12	Proportion of energy consumption powered by clean energy	%	50	60
	13	Forest coverage	%	41.2	≥ 41.2
	14	Greenway mileage per unit area	Kilometer/square kilometer	1.0	≥ 1.0
	15	Per capita park green space	Square meter/per person	16.9	17.4
Low-Carbon Environment	16	R&D investment as a proportion of GDP	%	4.0	4.5
	17	Low-carbon technology investment as a proportion of total R&D investment	%	10	15
	18	Carbon Emission Statistics, Accounting and Assessment System	—	Established	Advanced Practice
	19	Citizens' awareness of low-carbon ideas	%	80	90

Notes: Carbon dioxide emissions per unit of GDP and energy consumption indicators per unit of industrial added value are calculated relative to the 2005 baseline. The indicators for 2015 are based on the Shenzhen National Economic and Social Development 12th Five-Year Plan and the special plans of various departments. The numbers for 2020 are forecasts. The assessment indicators are adjusted according to the actual implementation of the 13th Five-Year Plan for National Economic and Social Development of Shenzhen Municipality.

Source: Adapted by author from Shenzhen Municipal People's Government (2012).

The Leadership Approach to Low-Carbon Policy Experimentation

The bridge leader in Shenzhen, Vice Mayor Tang, was one of the strongest local advocates of low-carbon policy experimentation among the four case studies. He worked with Mayor Qing Xu from 2009 to 2015. Mayor Xu had been a national official before he was assigned as a top politician in Shenzhen (starting as the Party secretary of Shenzhen municipal government). Tang, by contrast, was a professor of economics before serving twenty-one years as a local official in Shenzhen, initially as deputy director of the General Office of Shenzhen municipal government, and later in the Shenzhen Municipal People's Congress. Tang was instrumental in all steps, from mobilizing political support and local implementation alliances to promoting low-carbon policy programs and grounding the institutional foundations for Shenzhen's long-term decarbonization.

Shenzhen emissions trading system

Without Tang's leadership, Shenzhen would not have been a national ETS pilot in 2011. Due to an administrative delay, the national pilot list was nearly finalized before he became aware of the opportunity.[2] After learning that Shenzhen was not on the list, Tang began to mobilize political support for engaging in the project by reaching out to Mayor Xu. Xu then called the vice director of the National Development and Reform Commission, Xie Zhenhua, to express Shenzhen's willingness to become an ETS pilot, noting that it is the country's leading financial center (Personal Conversation 102000 2024; Interview 240107a 2013). Tang's ability to secure the mayor's support paved the way for Shenzhen's further engagement in emissions trading.

A wide range of local actors from bureaucratic agencies, local enterprises, and research institutes engaged in establishing the ETS pilot program. Professionals from bureaucratic agencies, research institutes, and other fields participated in the research detailing the design of the Shenzhen ETS, proving Shenzhen was a credible case for consideration as a national ETS pilot (Interview 232407 2013; Interview 240107a 2013).

In 2013, Tang led a group of thirteen local professionals with diverse backgrounds in investment, finance, energy, urban development, local enterprises, and research institutes on a nine-day visit to Belgium, France, and Germany to study the European Union Emissions Trading Scheme (EU-ETS). They met more than thirty system designers and practitioners from leading agencies—including the European Commission Directorate-General for Climate Action; the European Energy Exchange; the German Federal Ministry for the Environment, Nature Conservation, and Nuclear Safety; and the German Emissions Trading Authority—and held extended dialogues on design, quota allocation, registration monitoring, and exchange procedures (Shenzhen Carbon Trading Study Tour 2013).

Upon their return, the team identified the key challenge of preparing the Shenzhen ETS—designing the quota allocation system. To prevent the market failure that occurred in the EU–ETS, in which the low demand for carbon quota exchanges led to sharp drops in carbon prices, at times to zero euros, the research team proposed a quota rescheduling system to give the municipal government the authority to adjust the quotas allocated to participants in case of market failure (Shenzhen Carbon Trading Study Tour 2013).

Another challenge faced by the local government as it prepared to launch the Shenzhen ETS was mobilizing local implementation alliances. With support from top political leaders, the local government created a leading group for Shenzhen ETS, with an executive office in the Shenzhen Development and Reform Commission, to coordinate the policy implementation with a variety of actors inside and outside the government. For example, the Shenzhen branches of the State Administration of Foreign Exchange and the People's Bank of China collaborated to make the Shenzhen ETS the first in the country to allow foreign investors to participate in the market with settlement in foreign currency or cross-border RMB. The Shenzhen Market Supervision Administration issued standards for measuring, reporting, and verifying industrial greenhouse gas emissions. It also helped industrial actors prepare their carbon emissions data to participate in the ETS (Shenzhen City Development Research Center and Shenzhen Emissions Exchange 2015).

Most local enterprises knew little about the operation of the ETS, which made them reluctant to participate (Interview 010712a 2013). Other pilots, including those in Beijing and Shanghai, faced similar challenges. But Tang began to engage in dialogues with the standing committee of the Shenzhen Municipal People's Congress and the Shenzhen municipal government to promote the idea of codifying emissions trading regulations into local legislation. An expert in carbon business consulting described, in an interview for this study, how Tang successfully conveyed the urgency of the ETS pilot to the Shenzhen Municipal People's Congress by mobilizing his contacts and networks from his time as the vice director of the committee of the Municipal People's Congress (Interview 240107a 2013). The provisions on managing Shenzhen emissions trading passed through the Shenzhen Municipal People's Congress in 2012, which gave the municipal government the authority to request that local carbon-intensive enterprises join the ETS, and to determine the scope of the system's coverage and the quotas allocated to participating entities. Shenzhen's rapid embrace of legislation on its ETS program came about as a direct result of Tang's efforts.

In 2013, as a result of local leadership, Shenzhen become the first pilot to launch an ETS program, which more than six hundred industrial entities joined. Despite their initial hesitation, local enterprises were legally required to participate in the Shenzhen ETS (Interview 240107a 2013; Interview 181619 2013).

The launch of the Shenzhen ETS also led the city to institutionalize its low-carbon policy by creating legislation, regulations, standards, and a government division dedicated to low-carbon issues. Following the ETS legislation in 2012, the Shenzhen municipal government introduced a series of regulations and standards detailing the governing structure, quota allocation criteria, and measures to manage membership, exchange risk, handle violations and defaults, and exchange settlements to ensure the operation of the Shenzhen ETS (Shenzhen City Development Research Center and Shenzhen Emissions Exchange 2015). These measures allowed the Shenzhen ETS to continue to operate even after Tang and Xu left office, resulting in a 48.72 percent reduction in carbon intensity of the covered manufacturing industries between 2011 and 2018.[3] The Shenzhen ETS opened the country's first capacity-building training center in 2016.[4]

The Shenzhen International Low-Carbon City Program

Shenzhen's approach to low-carbon experimentation has been largely driven by its government leadership. Pingdi, a sub-district of Longgang District, was a low-income, high-carbon emissions area with a carbon intensity 2.5 times the city's average (Shenzhen Institute of Building Research 2013, 11). The Shenzhen International Low-Carbon City Program was originally created from the Sino–Dutch (Europe) Low-Carbon City project, which was designed to resolve the regional development gap in Pingdi and its neighboring districts by developing them as a low-carbon corridor in eastern Shenzhen (General Office of Shenzhen Municipal Committee of the Communist Party of China 2012). The Sino–Dutch project aimed to achieve rapid economic growth while gradually reducing the intensity of carbon emissions to the level of EU countries' 2010 average. Government leaders in Shenzhen reached out to the national government for resources and guidance on how to be a national model for other cities in similar transitions. On December 31,

2011, the Sino–Dutch (Europe) Low-Carbon City project was renamed the Shenzhen International Low-Carbon City Program and became part of the China–European Union partnership on climate change.

A low-carbon solution to a development challenge

The Sino–Dutch (Europe) Low-Carbon City project was launched as the Shenzhen government sought to forge new growth opportunities for the economically lagging area of Longgang. The local government reached out to its Dutch counterparts for policy ideas and international collaboration. The project was touted predominantly as a local project in collaboration with the Netherlands.

A research group led by Tang and Ton van Zeeland (Consul General of the Netherlands Consulate in Guangzhou) was established and funded by the Shenzhen municipal government, the Longgang district government, and the Next Generation Infrastructures Foundation (the Netherlands) to study the planning, concepts, and projects associated with low-carbon development (Wang et al. 2012). The group went on a study tour of the Netherlands from September 19–23, 2011, which included visits to multiple cities as well as bilateral meetings with the Ministry of Infrastructure and the Environment as well as the Ministry of Agriculture, Nature, and Food Quality, both of which supported the idea behind the project.

After their return, the local government leadership in Shenzhen sought to integrate the project into the bilateral initiatives between China and the Netherlands to make it a national flagship program on low-carbon city development (General Office of Shenzhen Municipal People's Government 2011). The Shenzhen government arranged a Sino–Dutch (Europe) Low-Carbon City Seminar at a time that coincided with the Netherlands' ministry-level visit to China (Development and Reform Commission of Shenzhen Municipality 2011b).[5]

The Shenzhen government leadership reached out to the National Development and Reform Commission to promote the idea of the Sino–Dutch (Europe) Low-Carbon City and expressed their willingness to scale up to a European Union–China bilateral project on climate change coordinated at the national level. In October 2011, the Shenzhen government made three requests for central government support in attracting international attention. First, it asked the central government to list the project as a discussion topic during the visit of the Netherlands' Ministry of Infrastructure and the Environment to the National Development and Reform Commission in November 2011 and to sign a ministry-level cooperation framework. Second, Shenzhen asked to integrate its low-carbon city project into the EU–China Partnership on Climate Change and to sign an initial cooperation agreement with the European Union at the Durban Climate Conference.[6] Third, city officials invited the National Development and Reform Commission to the Sino–Dutch (Europe) Low-Carbon City Seminar, held in November 2011, to provide guidance on its cooperative model, basic infrastructure, technology, and local carbon research for the development of local industries (Development and Reform Commission of Shenzhen Municipality 2011a).

The local government leadership took advantage of the policy and networking opportunities to integrate local low-carbon initiatives with national and international climate policy actions. Becoming a high-profile, national-level project would enable the local low-carbon program to access further international and national resources in the implementation process. The Sino–Dutch (Europe) Low-Carbon City Seminar involved more than eighty attendees from the government, academia, and enterprises from China and the Netherlands (Development and Reform Commission of Shenzhen Municipality 2011b).

The seminar generated a proposal to cooperate on urban planning, energy, green building, knowledge infrastructure, and governance. Both sides were interested in using the city of Eindhoven as a model for Pingdi's low-carbon construction plan and supported the idea of creating a working mechanism at the vice premier level between the Netherlands and China to coordinate implementation. In addition, local carbon emissions caps and an indicator system were proposed as part of China's low-carbon urbanization standards (Development and Reform Commission of Shenzhen Municipality 2011b). The exchanges with the Netherlands augmented Shenzhen's interest in the low-carbon city concept.

Pushing a local project to the global level

The Shenzhen government systematically sought external connections at the national and international levels to tap potential resources and bring the local project onto the global stage of low-carbon practices. The Shenzhen municipal government requested national government resources to integrate the Sino–Dutch (Europe) Low-Carbon City project into the higher-level government's agenda. It also asked the National Development and Reform Commission to consider the Sino–Dutch (Europe) Low-Carbon City project an "experimental point" for China's low-carbon development strategy with the goal of making local economic growth less energy intensive and more environmentally friendly. In 2011, the National Development and Reform Commission recognized Shenzhen's efforts and suggested further broadening its international cooperation with Germany, France, the United Kingdom, Denmark, Japan, Australia, and some US states and cities (Development and Reform Commission of Shenzhen Municipality 2011c).

The bridging leadership led by Tang helped Shenzhen's International Low-Carbon City Program become part of the national-level climate policy agenda and venture onto the global stage. Shenzhen leveraged its national and international connections to improve its access to international best practices and to attract investments and high-tech and low-carbon industries that would ultimately support its low-carbon transition (General Office of Shenzhen Municipal Committee of the Communist Party of China 2012). For instance, Dutch companies have provided technical assistance in building thc Distribution Energy Center and developing the pilot program's carbon emissions monitoring mechanism. German companies have also invested in developing new industries within the project (Cheshmehzangia et al. 2018, 71–72).

The low-carbon targets for 2015 included more than doubling the GDP and all new buildings meeting green standards. By 2020, the targets were much higher: the GDP was to be thirty times higher than 2010 levels (reaching 100 billion RMB), all new buildings would be green, 90 percent of transport would be green, 100 percent of energy would be clean, and the project area would be among the lowest-carbon urban areas in the world (General Office of Shenzhen Municipal Committee of the Communist Party of China 2012).

To achieve these low-carbon targets, all local industries were required to be evaluated according to the Catalogue of Industrial Development Guidance on the Shenzhen International Low-Carbon City, which encouraged six industrial areas: energy saving and environmental protection industries, low-carbon new energy industries, life and health industries, low-carbon service industries, high-tech equipment manufacturing industries, and aerospace and aviation industries (Shenzhen Institute of Building Research 2013, 30). One of the program's major strategies was to encourage large-scale, low-carbon projects to relocate to the project area, thus attracting low-carbon industries, project investment, and cooperation from international actors (General Office of Shenzhen Municipal Committee of the Communist Party of China 2012).

Resource Mobilization

The formation of the Leading Group for the Shenzhen International Low-Carbon City Plan and Construction to orchestrate the project's implementation in August 2012 proved instrumental to its resource mobilization (General Office of Shenzhen Municipal People's Government 2012c). This group, chaired by Tang, consisted of bureaucratic leaders from key municipal government administrative agencies, leading township government officials, and participants from local government-affiliated financial and investment enterprises. The bureaucratic agencies specialized in local policy planning, the local government-affiliated financial and investment enterprises mobilized capital and land for new construction, and the township government supervised the work to ensure it complied with low-carbon standards (General Office of Shenzhen Municipal People's Government 2012c).

Tang regularly organized group meetings to coordinate the implementation process. During these meetings, informal institutional connections were created to mobilize resources for policy implementation. The municipal government founded the state-owned company Shenzhen SEZ Construction and Development Group, which has been one of the major financing platforms for large urban construction projects since 2011. The municipal government appointed the company to manage the overall development and construction plan, and the company asked the Shenzhen Development and Reform Commission to help it evaluate land assets and program construction costs (General Office of Shenzhen Municipal People's Government 2012d). The municipal government allocated 330,000 m^2 of state-owned land to the company. It placed the Shenzhen Institute of Building Research in charge of the design and operational management of several of the program's construction projects (General Office of Shenzhen Municipal People's Government 2012b). It also assigned the Shenzhen Institute of Building Research, along with other local research institutes including the Shenzhen Water Planning and Design Institute and the Shenzhen Beilinyuan Landscape and Architecture Planning and Design Institute, to design the construction plan for building Pingdi as an international low-carbon city (General Office of Shenzhen Municipal People's Government 2012a). The annual International Low-Carbon City Forum in Shenzhen attracted several thousand guests from over forty countries and regions in 2016.[7]

Shenzhen has continued to support the development of low-carbon initiatives and institutions since the vice mayor retired in 2016 and the mayor left office in 2017. Similar to Zhenjiang, Shenzhen has created a Division of

Climate Change under the Shenzhen Ecology and Environment Bureau with specialized members of staff to coordinate climate change policies and low-carbon development. In 2020, the Shenzhen Municipal People's Congress passed the country's first local legislation on green finance, which encourages financial institutions to review the environmental performance of investments in Shenzhen based on their level of carbon emissions and incorporates the Shenzhen ETS as a managing entity in promoting domestic and cross-border transactions of carbon assets and green assets. In 2021, the Shenzhen government issued a new plan to drastically decarbonize the city's transportation system, with the goal of increasing the number of new energy vehicles to around one million by 2025 by ensuring that 60 percent of new registered vehicles between 2021 and 2025 are new energy vehicles (Development and Reform Commission of Shenzhen Municipality 2020). Shenzhen has built a reputation for decarbonizing urban transportation and has become one of the world's leading cities in climate action and urban sustainability (Li et al. 2020).

In summary, of the four case studies, Shenzhen has engaged the most substantively in its low-carbon policy experiment. The municipal government began to advocate decarbonization after its leaders cautioned that the city's heavy reliance on the consumption of natural resources was unsustainable. This case also demonstrates the important role of the bridge leader in the process of fostering low-carbon policy institutionalization that is essential to shape the direction and outcome of city's efforts toward decarbonization.

5
Performative Engagement
Nanchang

Nanchang, the capital city of Jiangxi Province, was a somewhat unlikely early advocate of low-carbon development. Many Chinese people consider it a lagging inland city. Yet Nanchang was one of the first Chinese cities to experiment with the idea of low-carbon development, even before the launch of the national low-carbon pilot city program. As a pilot city, it pledged to be a model for exploring paths toward low-carbon development for inland cities. A mid-level bureaucrat—the chief economist—played an important role in proposing policies to reduce carbon emissions and building implementation alliances with key local actors, such as the local people's congress, vice mayor, higher and lower levels of government, and international agencies. Nanchang's major low-carbon policy initiatives included decarbonizing the transportation and building sectors as well as greening the local government evaluation system. It also passed one of the first local laws in China to promote low-carbon development in an effort to gain national government recognition but lacked enforcement. This chapter examines Nanchang's performative engagement in low-carbon policy experimentation from 2010 to 2018.

Unlikely Forerunners in Low-Carbon Policy Experiments

When the National Development and Reform Commission selected Nanchang as a low-carbon pilot city in 2010, its motivation to engage in low-carbon economic development was compelling. The global economic crisis of 2009 pushed the local government to devise a development strategy to make local growth less dependent on increased energy and resource consumption (Interview 081200a 2013). At the time, Nanchang had few large-scale industrial enterprises; its economy mainly relied on uncompetitive local enterprises: fewer than 10 percent engaged in technical innovations

Implementing a Low-Carbon Future. Weila Gong, Oxford University Press. © Oxford University Press (2025).
DOI: 10.1093/9780197757451.003.0005

and were unlikely to help achieve the local government's goal of catching up with other areas in terms of development (Li and Xiong 2010). In the early 2010s, the city was eager to foster a new branch of high-tech industries and decarbonize its energy mix (see Table 5.1). The local government hoped to catch up economically through low-carbon development. As its chief economist noted, "As a developing city, Nanchang will face challenges in keeping up with other Chinese cities if it continues to follow traditional growth patterns. However, given its strong ecological foundation, the city should explore how to leverage its environmental advantages to pursue green development—and potentially surpass faster-growing cities."[1]

Nanchang's energy intensity sharply decreased after it became a low-carbon pilot city, from 0.45 in 2010 to 0.34 tonnes of coal equivalent per GDP unit in 2015. The 46.4 percent reduction in energy intensity between 2005 and 2015 surpassed its target of 38 percent, as stipulated in its low-carbon pilot action plan (see Table 5.1) (Nanchang Development and Reform Commission 2010; Xiong et al. 2020, 27). However, its total energy consumption from 2010 to 2018 continued to rise (see Figure 5.1) (Jiangxi Provincial Energy Bureau 2020).

Table 5.1 Nanchang's Low-Carbon Pilot City: Goals

No.	Aim	Unit	2010	2015	2020
General Goals					
1	Carbon emissions reduction per GDP compared to the 2005 level	%	25	38	45–48
Industry Structure Modification Goals					
2	High-tech value-added as a percentage of industries above designated size	%	29	35	45
3	Service industry as a proportion of GDP	%	39	43	48–50
Energy Mix Modification Goals					
4	Non-fossil energy as a percentage of primary energy utilization	%	3	7	15
Others					
5	Forest coverage	%	21	24	28
6	Living wood growing stock	10,000 m^3	240	380	420

Source: Nanchang Development and Reform Commission (2010).

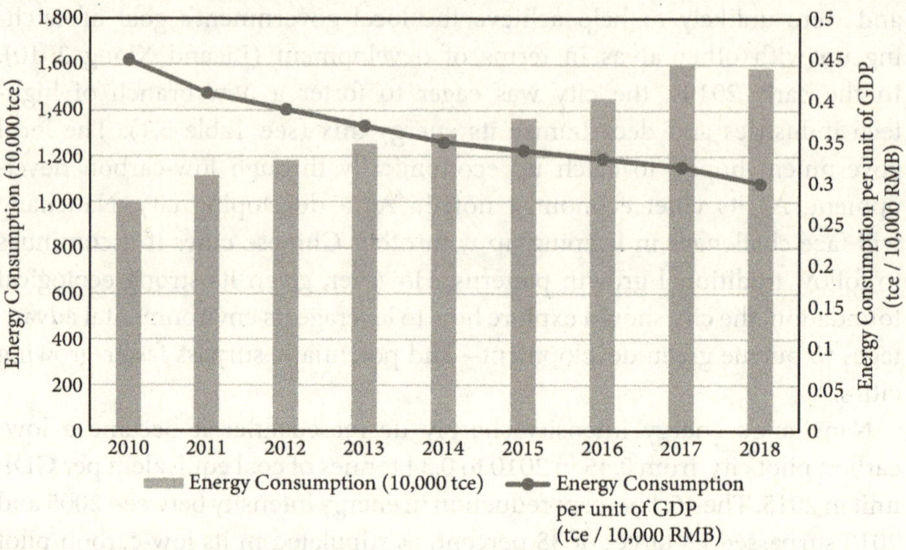

Figure 5.1 Nanchang's Energy Consumption and Energy Consumption per unit of GDP (2010–2018)

Note: Data on Nanchang's total energy consumption and energy consumption per unit of GDP for the years 2010–2016 are cited from Xiong et al. (2020). Data for 2017 and 2018 are based on the author's calculations based on data from Xiong et al. (2020) and the Statistical Communiqué on National Economic and Social Development of Nanchang for the respective years.
Source: Xiong et al. (2020); Statistics Bureau of Nanchang Municipality and Survey Office of the National Bureau of Statistics in Nanchang (2018, 2019).

The Leadership Path toward Low-Carbon Policy Experimentation

Yong Shang, a member of the politburo and the deputy Party secretary of Jiangxi provincial government from 2008 to 2014, engaged Nanchang in low-carbon activities well before it became a national pilot city. Shang has a PhD in engineering; he promoted energy efficiency and reducing carbon emissions when he worked as a vice minister of science and technology. He chaired the World Low-Carbon and Eco-Economy Conference and Exposition, held in the city three times between 2009 and 2014, before he left Jiangxi (Nanchang Convention and Exhibition Office 2009)

In 2009, the conference involved sixteen government departments and working units, including the National Development and Reform Commission, the Ministry of Science and Technology, the Ministry of Housing and Urban–Rural Development, and the Ministry of Finance. More than thirty projects were finalized at this conference, worth more than 50 billion RMB

(US$7 billion) (World Low-Carbon and Eco-Economy Conference, 2009).[2] In the conference, Nanchang became the first Chinese city to call for worldwide efforts to advance low-carbon development, and to address the environmental challenges associated with climate change and resource and energy use.

Nanchang's early involvement with low-carbon issues prepared it to grasp the opportunity to become a national pilot. In 2010, it created a Leading Group for Low-Carbon Pilot City Work within the municipal government. The Nanchang Development and Reform Commission has since taken over the role of coordinating the resource mobilization and policy implementation process (National Center for Climate Change Strategy and International Cooperation 2017, 171). Shang's departure did not affect the city's continued participation in decarbonization activities. The bridge leader—Chief Economist Hua Liu—from the Nanchang Development and Reform Commission continued to push low-carbon issues on the government policy agenda.

Chief Economist Liu as the bridge leader

In contrast to Shenzhen's experience, which highlights the importance of securing political support from the top local political leader to facilitate sweeping policy progress, Chief Economist Liu pursued a very different approach: he used his personal authority to push the low-carbon issue onto the local policy agenda without the direct involvement of high-profile political leaders.

Liu was a leading figure in Nanchang's low-carbon policy experimentation; he promoted low-carbon policies in the city even before the establishment of the Jiangxi Climate Change Division in 2009. Liu joined the Nanchang Planning Bureau (later renamed the Nanchang Development and Reform Commission) in 1989. By the time he left, in 2017, he had risen to become a chief economist and a member of the Commission's Party Committee, which wields decision-making power within the commission. He was able to influence the municipal development planning process as well as funding and project applications. He laid the groundwork for Nanchang's low-carbon policy initiatives and updated the Department of Climate Change at the national level on the city's low-carbon policy progress (Interview 061900 2018).

Even without the full support of top municipal government leaders, Liu was able to reach out to higher levels of government and non-government actors to mobilize resources in the form of trained personnel for policy implementation. Under Liu's leadership, the Nanchang Development and Reform Commission organized low-carbon research and meetings with external research institutes, such as the Chinese Academy of Social Sciences (Interview 02232426x 2018). After an introduction from the National Development and Reform Commission, Nanchang also engaged in low-carbon projects with Austria's Transportation and Technology Ministry (Interview 02232426x 2018). Between 2010 and 2011, more than two hundred local officials received low-carbon training through China–Austria low-carbon forums and study tours to Austria. Nanchang also worked with the British Council to organize training sessions for local enterprises during this time (National Center for Climate Change Strategy and International Cooperation 2017, 171).

The vice mayor officially headed the Nanchang Leading Group for Low-Carbon Pilot City Work, created in 2010, which comprised dozens of division leaders from multiple bureaucratic agencies in the municipal government. Yet, in practice, Liu and two of his colleagues in the Nanchang Development and Reform Commission ran the city's low-carbon pilot work. One colleague had worked with Liu since the start of the pilot, and the other joined in 2013. Although the political leaders of the municipal government and the Nanchang Development and Reform Commission changed several times during nearly a decade of low-carbon policy experimentation, the policy process was not affected much until Liu was charged with corruption and taken into custody in 2017 (Interview 100824 2018; Interview 061900 2018; Interview 081724b 2018). As illustrated in the following discussion of how the local bridging leadership created implementation alliances for policy implementation, Liu pursued a two-pronged approach to low-carbon development that relied on his strategic position and the exceptional personal authority he accumulated over decades of work on the Commission.

Resource Mobilization

Cities need leaders who are willing to drive local low-carbon policy experimentation. The mobilization strategies they use to form implementation alliances in the policy process are equally important. Although several researchers have attributed local governments' ability to mobilize support for policy innovations mainly to local leaders' personal characteristics and personal networks (Hasmath et al. 2020; Teets 2018; Tie et al. 2020), Nanchang's low-carbon policy experiments did not entirely rely on individual leaders' personal authority and connections, but also on their mobilization strategies.

Metro construction

As in many other Chinese cities, the fast growth of private vehicles created heavy traffic jams in Nanchang (World Bank 2012). Decarbonizing the transportation sector is thus one of the key issue areas on Nanchang's low-carbon policy agenda (see Table 5.2). Despite earlier debates among various

Table 5.2 Nanchang's Low-Carbon Pilot City: Key Projects

Categories	Contents	Amount	Investment scale (billion RMB)
1. Solar PV	Construction and promotion of photovoltaic power sources and application	8	6.285
2. LED	High-power high-brightness, LED epitaxy and chip production, LED industrial park construction, LED lamp promotion and application	4	10.868
3. Service outsourcing	Software industrial park construction	3	2.571
4. Cultural tourism	Culture and creative design, ecological tourism park construction	9	16.095
5. Low-carbon technology service platform	Construction of thin film solar cell research and development (R&D) center, Transparent conducting oxide glass R&D center, crop seed breeding base, forestry system database and training network	4	0.667
6. Low-carbon transportation	Metro Lines 1 and 2 phase I construction, development of public bicycle program	2	33.1
7. Other	New energy vehicles, natural gas utilization, biogas utilization, afforestation	22	12.153
Total		**52**	**81.739**

Source: Nanchang Municipal People's Government (2011).

bureaucratic agencies on the feasibility of building a metro system given the city's relatively constrained local financial capacity, the municipal government placed metro construction on the local policy agenda in 2008 with the aim of increasing the capacity of public transportation and reducing traffic congestion. Nanchang expanded the length of its metro system threefold, from 28.8 km in 2016 to 88.8 km in 2020, and aimed to more than double its coverage to 200 km by 2025. The goal was for the completed metro to carry an average of 1.3 million passengers per day and provide half of the city's public transport needs (Nanchang Metro 2021).

Nanchang attracted the support of higher levels of government and international sources to fund the construction of its metro. The national government approved the city's metro construction proposal. This project entailed a total investment of 15.974 billion RMB (US$2.34 billion) by 2019, including US$250 million in 2013 from the World Bank as part of its Country Partnership Strategy (2011–2015) with China. Domestic bank loans provided the rest of the funding for the project (Jiangxi Provincial Audit Office of the People's Republic of China 2017). The Nanchang municipal government also received one billion dollars in bonds from the Jiangxi provincial government in 2019, 70 percent of which was for the construction of urban rail Line 1.[3] Thus, even though it was an economically lagging area, the municipal government was able to mobilize external funding from higher levels of government and international organizations to decarbonize its public transportation sector.

Greening local government performance evaluation criteria

Even before Nanchang's engagement in the national low-carbon pilot city program, the municipal government attempted to introduce new incentives for the local government to transition local growth away from a heavy reliance on the consumption of natural and energy resources. In 2008, the municipal government issued an executive order to introduce environmental metrics into local government performance evaluations to encourage township governments to green development by not primarily on GDP growth; the evaluations also considered the impact on natural and material resources. As a local official working for a township government recalled, "The municipal government removed the industries' output growth, added forest coverage, water sources and quality, and ecological tourists as indicators into the government performance evaluation system. The weight of ecological achievement indicators accounted for almost one-third in the overall assessment."[4] Wanli township in Nanchang was selected to trial the new evaluation system. In

2012, its municipal government announced it would "keep on promoting Wanli environmental protection and scientific development," which refers to the allocation of millions of RMB to subsidize its environmental and ecological protection, and tourism development efforts, among others (General Office of Nanchang Municipal Government 2012).

Yet, lower-level governments did not always support new policy measures introduced by higher-level governments. A Wanli government officer noted that township governments were cautious about implementing the new evaluation measures because local leaders believed their areas deserved an equal chance of economic development and felt the performance evaluation system disadvantaged their towns (Interview 180312a 2013). The source further commented that the new evaluation system did not seem to be very effective, even though polluting industries did move out of town. Emissions had risen as construction projects increased in the central township area (Interview 180312a 2013). Therefore, even in a centralized political system like China's, policy experiments proposed by a higher level of government are not always implemented at the lower level. Implementation challenges remained, particularly when allied actions could not be forged across different levels of government.

Building energy efficiency and green building regulations

A key component of Nanchang's low-carbon pilot action plan is to promote energy efficiency and the construction of green buildings—for which the Nanchang Housing and Urban–Rural Development Bureau was responsible well before the launch of the pilot city program. An employee of the bureau noted that a lack of resources and political attention in the policy process has constrained the city's development of building energy efficiency and green building (Interview 250808b 2018). In 2010, a new bureaucratic agency—the Building Energy Efficiency and Technology Division—was tasked with pushing green building development. The division is equipped with trained officials who specialize in promoting the use of environmentally friendly materials and supervising the implementation of national energy efficiency standards in the design and construction process.

Since the early 2000s, land concessions and the real estate industry became an important tax source for Chinese cities (Tian 2018; World Bank and the Development Research Center of the State Council 2014). According to a Nanchang government official interviewed for this research, some real estate developers who had close connections with the municipal and provincial governments complained that green building was too expensive: a project might incur an extra 100 million RMB in costs to satisfy green regulations (Interview 242612 2018). This led the vice mayor to question the responsible local officials about why green building was so expensive and placed such a huge burden on enterprises, and about the need to promote green building. The division head repeatedly explained to the vice mayor that real estate developers were exaggerating: the green building standards cost only an extra 20 RMB per 1 m^2, which represented a fraction of the cost of a development project (Interview 242612 2018).

As a result of the division head's persistent efforts to persuade the vice mayor who was overseeing the building sector in Nanchang, the city was able to move ahead with its building energy efficiency and green building programs. The region's green building development has rapidly grown. In 2010, the city had only one green building project that was accredited with national green building standards, which was a total of 60,000 m^2 in size; as of 2018 it had more than ninety projects with green building credentials, with a total size of over 130 million m^2 (Interview 242612 2018). The strengthening of the green building regulations in 2017 expanded the scope of green building applications from mostly government-administered and large public buildings (over 10,000 m^2) to mostly newly renovated or built public and civil buildings (Nanchang Housing and Urban–Rural Development Bureau and Nanchang Development and Reform Commission 2017). How can the Building Energy Efficiency and Technology Division create the necessary support to enhance its legitimacy and win over potential opponents in policy implementation?

Despite the challenges associated with securing higher-level political support to green building development, the Nanchang Housing and Urban–Rural Development Bureau demonstrated its ability to maintain local efforts to decarbonize the building and construction sector in four ways: elevating local expertise on green building; working with influential actors inside and outside the government to integrate green building standards into the local policy system; securing national government projects to obtain local matching funding; and persuading local leaders of the benefits of acting on low-carbon issues.

First, as Jiangxi's capital city, Nanchang is home to some of the province's major universities and research institutes. Thus, in 2012, the Nanchang Housing and Urban–Rural Development Bureau created a pool of local experts, including twenty-three specialists from local research institutes and technical companies. By engaging with local experts on drafting plans and standards, the Bureau was able to issue a series of planning and administrative regulations, such as the Nanchang Green Building Medium-Long Term Plan (2012–2020), the Nanchang Building Market Oversight Regulation, and the Nanchang Green Building Development and Oversight Implementing Regulations (2017–2020) (Nanchang Housing and Urban–Rural Development Bureau and Nanchang Development and Reform Commission 2012; Nanchang Municipal People's Congress 2015; Nanchang Housing and Urban–Rural Development Bureau and Nanchang Development and Reform Commission 2017).

Jiangxi began to issue green building evaluation standards due to Nanchang's fast development: 80 percent of the province's green building development took place there. As an officer from the Nanchang Housing and Urban–Rural Development Bureau explained, the Bureau leveraged experts in Nanchang and actively requested the promotion of green building standards. The Jiangxi provincial government issued related regulations in 2015 to meet the needs of the city's rapidly developing green building industry (Interview 242612 2018). Nanchang's experience demonstrates that even in a centralized political system, local governments can leverage expertise gained from administrating current programs to influence the policy process at higher levels of government.

The second way in which the Nanchang Housing and Urban–Rural Development Bureau demonstrated its ability to form implementation alliances with competent bureaucratic agencies was to strengthen green building regulations. In the Nanchang Green Building Medium-Long Term Plan (2012–2020), the Nanchang Housing and Urban–Rural Development Bureau justify that the development of green building regulations was in line with the city's low-carbon development goals, which were pursued by the Nanchang Development and Reform Commission (Nanchang Housing and Urban–Rural Development Bureau and Nanchang Development and Reform Commission 2012). The following year, the two bureaucratic agencies jointly issued the Nanchang Green Building Development and Oversight Implementing Regulations (2017–2020), which detailed systemic supervision and management rules designed to enforce the green building standards throughout the project process, from set-up to inspection and operation (Nanchang Housing and Urban–Rural Development Bureau and Nanchang Development and Reform Commission 2017). After this administrative order was issued, the development of green building accelerated in Nanchang. For instance, fifty-nine building projects, equal to 702 m^2, were approved as green building projects over the next two years.[5]

The Nanchang Municipal People's Congress passed the Nanchang Building Market Oversight Regulation in 2015, which made building energy efficiency and green building standards compulsory for all new government investments and commercial projects (Nanchang Municipal People's Congress 2015). This regulation also authorized the Nanchang Housing and Urban–Rural Development Bureau to closely inspect the implementation of civil construction projects through final check-ups on the energy efficiency outcomes and implementation of green building designs. The strengthening of green regulations has driven the subsequent spike in green building development in Nanchang (Interview 250808b 2018). Nanchang has become a leader in Jiangxi on green building development. As of 2019, over 95 percent of new building areas were designed according to green building standards, and more than 63 percent of new buildings complied with these standards (Jiangxi Housing and Urban–Rural Development Commission 2020).

Local legislation on low-carbon development

Nanchang was one of the first Chinese cities to push a local legislation on low-carbon development—Regulations on the Promotion of Low-Carbon Development of Nanjing City—through the municipal and provincial people's congresses. A key element of this legislation, from 2016, is that it requested local actions to institutionalize the low-carbon development plan and evaluations, and it required cross-sectoral collaboration to achieve the city's greenhouse emissions reduction goals. The legislation authorized the earmarking of annual financial support for low-carbon-related research and business development in Nanchang (Jiangxi Provincial People's Congress 2017).

The city's strong interest in passing the legislation coincided with local leaders' strong concerns about gaining national political recognition for local policy achievements. An official involved in the policy process noted that Nanchang could not have met the national government's expectations for its performance on low-carbon policy experimentation without taking pioneering steps in experimental legislative measures for decarbonization (Interview 081200b 2018).

When the Nanchang Development and Reform Commission proposed introducing legislation to promote low-carbon development in 2014, the Legal Affairs Committee of the Nanchang Municipal People's Congress challenged the Commission about the need for legislation. The Committee argued that if the legislation were promotional in nature, department rules were sufficient for the purpose. But, according to the Nanchang Development and Reform Commission, as it was unrealistic to make the regulatory requirements from the promotional legislation legally binding, promotion should remain the primary purpose (Interview 02232426x 2018; National Center for Climate Change Strategy and International Cooperation 2017, 171). The turning point in this lengthy debate was when the vice mayor helped the legislation move to the top of the legislative agenda and later helped push it through the Nanchang Municipal People's Congress in 2016 (Interview 02232426x 2018).

After months of discussions between the two agencies, the Legal Affairs Committee finally ensured that the legislation included a clause outlining the legal responsibilities of related entities (e.g., administrators, enterprise, and business owners). The behavior of these entities was to be regulated through mandatory carbon emissions reporting; violations could incur financial or administrative punishments. The promotional legislation encouraged related government agencies to incorporate low-carbon energy goals into their local economic development, urban construction, and consumer consumption policies. It also allowed the Nanchang Development and Reform Commission to integrate low-carbon policies into their drafts of major municipal government documents such as the socioeconomic development plan, urban–rural development master plan, and land use master plan. Nanchang's legislative efforts attained considerable provincial and national visibility and represent progress toward achieving the city's low-carbon aspirations.

However, the legislation's overall impact has been largely performative; once passed, the legislation was difficult to implement (Interview 081200b 2018). The initial flurry of national interest that followed its passage, in the form of providing additional financial support and expecting Nanchang to introduce further supplementary low-carbon policies, stalled in 2017 (Interview 081200b 2018).

Promoting policy experiments requires bridging leadership, which Nanchang had in abundance when Liu was there. But his alleged corruption reflects the flawed institutions that fused public and private interests. Low-carbon policy progress in Nanchang has slowed since Liu was arrested on corruption charges in 2017 (Interview 081724b 2018; Interview 061900 2018). The Nanchang Low-Carbon Development Association, a non-governmental organization, was later accused of being linked to Liu's corruption. Liu allegedly established the association as a front to facilitate bribes and money laundering.[6] After that, the Nanchang Development and Reform Commission enacted dozens of administrative reforms to address the institutional flaws identified in the management of the funding, bidding, and project approval processes.[7]

Nanchang's subsequent lack of low-carbon policy enforcement was mainly attributed to a shortage of trained personnel to oversee implementation. Given local officials' heavy workloads, they had little time or energy to advance low-carbon initiatives when these were not prioritized by the government or high on the agenda of their bureaucratic divisions. After Liu left, only one person was assinged to low-carbon tasks, and only on a part-time basis (Interview 02232426y 2018). Until then, the low-carbon financial fund had not been used because the inventory for low-carbon technology had not been issued, which prevented the start of the fund application process (Interview 02232426y 2018). In Shenzhen, a local low-carbon institution—the China (Shenzhen) Emission Exchange—was created and staffed with trained personnel to deal with greenhouse gas emission and energy conservation issues before the passage of the city's emissions exchange legislation. Yet Nanchang lacked a similar low-carbon institution with a full-time specialist staff to implement the proposed policies, jeopardizing its long-term low-carbon transition. The legislation also could be revoked if it is not well enforced, given the lack of specialized personnel (Interview 02232426y 2018).

Conclusion

Nanchang was a somewhat unexpected candidate for participating in the national low-carbon pilot city program. While not economically advanced, the city's initial low-carbon engagement began even before the national government launched the pilot program. The Nanchang municipal government viewed decarbonization as a strategic opportunity to help it catch up with economically advanced cities by modernizing its economic and industrial structure with low-carbon features.

Although it had insufficient political and financial resources, Nanchang was able to introduce a series of low-carbon initiatives in its transportation and building sectors. The municipal government mobilized abundant financial resources from national and international sources to build its underground railway. The city's bureaucratic agencies built cross-divisional alliances across different levels of government to promote regulatory measures in greening the building sector.

Chief Economist Liu, acting as the bridge leader, was the primary champion of Nanchang's low-carbon city policy experimentation. To gain national government recognition, he managed to build a cross-divisional alliance within the government to push one of the first pieces of local low-carbon legislation through the municipal and provincial people's congresses. However, the legislation has never been fully enforced. As of 2023, the city's specialized fund to foster low-carbon industrial development and innovation remained untouched, as no government policy guidance was issued to encourage industries to apply for this support—largely due to a lack of trained personnel to oversee its implementation after the bridge leader's departure in 2017.

6
Symbolic Engagement
Zhenjiang

Zhenjiang, located at the intersection of the Yangtze River and Jinghang Canal, is a port city in eastern China. While it is known for its national forest parks and historical sites, it is also a thermal power production center that supplies electricity to eastern China. Thus, coal is a major local energy source. However, Zhenjiang successfully branded itself as an active promoter of low-carbon initiatives soon after it became a national low-carbon pilot city. Despite changes in political leaders, the city has maintained its engagement in low-carbon initiatives—particularly those facilitated by the bridge leader in the Zhenjiang Development and Reform Commission. Through its participation in the national low-carbon pilot city program, bureaucratic agencies in the Zhenjiang municipal government have introduced several institutional initiatives to integrate low-carbon issues into existing governing and regulatory policy systems, including the creation of a low-carbon development division in the municipal government, incorporating green building indicators into the land approval process, as well as pushing low-carbon city standards to the provincial government level. The city's low-carbon engagement can remain symbolic if the participation of the industrial sector, a major contributor to the city's overall carbon emissions, stays limited. This chapter examines the implementation of low-carbon city policy experimentation in Zhenjiang from 2012 to 2018.

Local Challenges Related to Decarbonization

The city has experienced challenges in the transformation of its industrial structure. Energy-intensive industries involved in chemical, paper, and cement production have long been an important component of local economic growth, but they also degrade the local environmental quality. Zhenjiang's industrial structure is dominated by energy-intensive sectors such as

Implementing a Low-Carbon Future. Weila Gong, Oxford University Press. © Oxford University Press (2025).
DOI: 10.1093/9780197757451.003.0006

coal-fired power and chemical industries, which account for 70 percent of the city's total industrial energy consumption (Department of Ecology and Environment of Jiangsu Province 2020). At the start of its low-carbon policy experiments in 2012, carbon emissions from the power sector amounted to 28.82 million tonnes or 68.8 percent of those from the city's industrial sectors. Industrial sectors as a whole accounted for 73 percent of the city's total carbon emissions at that time (Liu et al. 2014).

One challenge associated with Zhenjiang's decarbonization is its status as a major power generation base for exporting electricity to other areas; roughly half of the power generated in Zhenjiang is exported (Zhenjiang Development and Reform Commission 2017b, 3). It is home to some of the largest coal-fired power plants in Jiangsu Province, some of which are owned by a state-owned company beyond the control of the Zhenjiang municipal government (Wang et al. 2015).

Given the impact of the industrial sector—particularly the power sector, which is not entirely under the control of the municipal government—it is not surprising that Zhenjiang's energy consumption remains high despite efforts to modernize its industrial structure. According to data from Zhenjiang Statistical Bureau, the city decreased the share of manufacturing and construction industries by 13 percent between 2010 and 2018, but energy consumption during that time did not significantly decrease (see Figure 6.1; Statistics Bureau of Zhenjiang Municipality and Survey Office of the National Bureau of Statistics in Zhenjiang 2019).

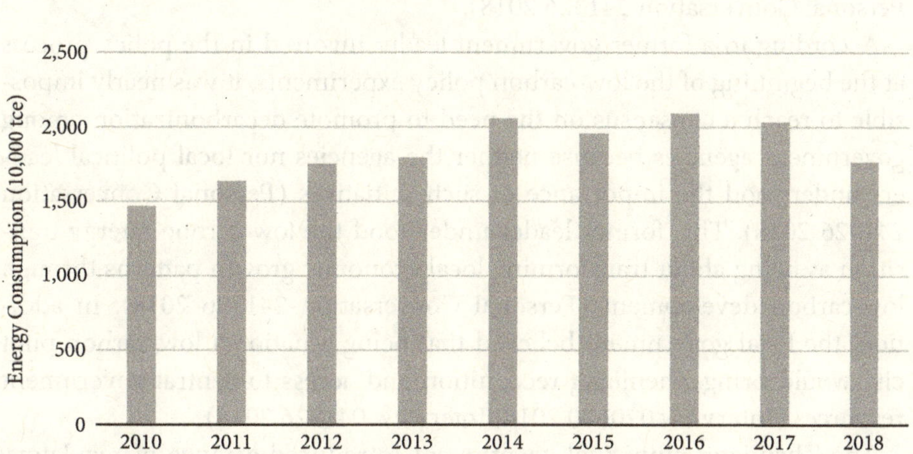

Figure 6.1 Zhenjiang's Industrial Energy Consumption (2010–2018)

Source: Statistics Bureau of Zhenjiang Municipality and Survey Office of the National Bureau of Statistics in Zhenjiang (2011–2019).

The Leadership Path toward Low-Carbon Policy Experimentation

Zhenjiang's low-carbon activities were initiated by the city's municipal government and led by local government leaders who were attuned to low-carbon policy experiments. Because it was very hard to push local low-carbon initiatives through the local government without the endorsement of authoritative figures (Personal Conversation 241326 2018), the Zhenjiang Leading Group for Low-Carbon City Development was formed and co-chaired by the top political leaders—the party secretary and the mayor. The involvement of top leaders signaled the local government's strong political support for pursuing low-carbon development (National Center for Climate Change Strategy and International Cooperation 2017, 5). The leading group urged the broad participation of various bureaucratic agencies by prioritizing the low-carbon issue on their bureaucratic agendas and local cadre evaluation criteria. The city also created a multi-layer leadership structure to coordinate low-carbon implementation between the municipal- and district-level governments. Political and bureaucratic leaders from related bureaucratic agencies, such as the municipal-level Urban Planning Bureau, Housing and Urban–Rural Development Bureau, Development and Reform Commission, Transportation Bureau, and Urban Management and Law Enforcement Bureau, and the district governments were responsible for implementing more than one hundred low-carbon tasks each year (Interview 171019 2018; Personal Conversation 241326 2018).

According to a former government leader involved in the policy process at the beginning of the low-carbon policy experiments, it was nearly impossible to reach a consensus on the need to promote decarbonization among government agencies because neither the agencies nor local political leaders understood the importance of such initiatives (Personal Conversation 241326 2018). This former leader understood the low-carbon energy transition as being about transforming local economic growth patterns through low-carbon development (Personal Conversation 241326 2018). In addition, the local government believed that being a national low-carbon pilot city would bring Zhenjiang recognition and access to central government resources (Interview 020600 2018; Interview 041826 2018).

The Zhenjiang municipal government introduced a range of regulatory measures to encourage low-carbon development. It also launched two high-profile policy initiatives. The first was a carbon management system designed to help the government collect data on (and monitor) major carbon emitters. The second was a carbon assessment project that sought to regulate the carbon emissions of all the city's existing and planned fixed asset investments.

In addition, it issued a number of low-carbon-related action plans ranging from a master plan on mid- to long-term low-carbon development to a specific action plan on building energy efficiency and low-carbon transportation (National Center for Climate Change Strategy and International Cooperation 2017, 297). Since President Xi Jinping recognized Zhenjiang's leading efforts in low-carbon construction and management during his visit in 2014, Zhenjiang has been invited to almost all of the high-profile international climate conferences on local practices (Zhenjiang Development and Reform Commission 2017b, 5).

Zhenjiang's low-carbon initiatives were built on the implementation alliances mobilized by the bridging leadership in the municipal government. The city's bridge leader was a deputy director of the Zhenjiang Development and Reform Commission. He became familiar with low-carbon issues while working as a staff member in the mayor's office. The deputy director played a central role in developing low-carbon city evaluation standards, prioritizing low-carbon issues on the commission's action agenda, and building alliances with higher-level government agencies, including the Jiangsu Quality and Technical Supervision Commission and the National Center for Climate Change Strategy and International Cooperation. The deputy director was able to work with the vice mayor, who continued to endorse low-carbon development at the municipal government level even after the new mayor expressed concerns about the effectiveness of low-carbon policy relevant to local development (Interview 020600 2018; Interview 041826 2018). For example, although it met the energy efficiency improvement target, Zhenjiang's economic performance was relatively weak—an annual growth rate of 3 percent, less than half the national level in 2018; it also missed its own goal for local economic development (Zhang 2019). Nevertheless, out of the low-carbon policy experimentation, Zhenjiang was able to maintain its engagement in low-carbon policy initiatives through a series of institutional efforts. For example, it created the Low-Carbon Development Division with full-time staff inside the government, promoted low-carbon city standards to the provincial level, and integrated green building standards into the local project approval process.

Resource Mobilization

Zhenjiang was one of the first low-carbon pilot cities to propose setting standards to evaluate the outcomes of low-carbon city initiatives. The city government has a national-level alliance, the National Center for Climate Change Strategy and International Cooperation, which helped solidify its reputation as a leading supporter of low-carbon policy experiments. This center, as the think tank affiliated with the Department of Climate Change under the Ministry of Ecology and Environment, has a leading role in shaping national-level climate change and low-carbon policies and facilitating collaboration and exchanges among domestic and international actors in decarbonization.[1] In 2016, the center released its first low-carbon city evaluation system, which it named the "Zhenjiang index" (National Center for Climate Change Strategy and International Cooperation 2016).

Low-Carbon City Evaluation System: Pushing regulatory measures to the provincial level

An entrepreneurial bureaucrat at the Energy Conservation and Environmental Protection Division in the Zhenjiang Development and Reform Commission originally proposed setting up an evaluation system for low-carbon city development. The proposed initiative quickly proceeded when the bureaucratic leader backed the proposition, particularly by reaching out to external actors to build implementation alliances. A local official involved in the policy process noted that he was personally motivated to explore a low-carbon city evaluation system to gain favor with his superior, a deputy director of the Zhenjiang Development and Reform Commission on low-carbon issues. Otherwise, he felt it would be meaningless for him to take such an initiative (Interview 082600 2018). Indeed, in local governments, without the support of bureaucratic leaders, staff-level officials lack the authority, resources, and policy entrepreneurship needed to implement low-carbon policy experiments.

The Zhenjiang Development and Reform Commission then contacted the city's Quality and Technical Supervision Bureau with the idea of setting up a low-carbon city evaluation system. Within a couple of months, the Zhenjiang government approved the proposal and set it as an annual prioritized task on the agenda of the Low-Carbon Development Division (in the Zhenjiang Development and Reform Commission). Then the Zhenjiang Development and Reform Commission, together with the Zhenjiang Quality and Technical Supervision Bureau, applied for project listing approval from the Jiangsu Provincial Quality and Technical Supervision Commission. After several months, the evaluation system was put onto the Commission's agenda (Interview 082600 2018; Zhenjiang Development and Reform Commission 2018).

The Zhenjiang Development and Reform Commission, together with the Jiangsu Quality and Technical Supervision Commission, then created a preparation team to research the design of a low-carbon city evaluation system. The team acknowledged that its system would be more applicable to cities with similar local socioeconomic conditions, such as those on the east coast rather than those in the middle to western areas, which are less economically advanced. One of the biggest challenges associated with designing the system was selecting the proper indicators (Interview 082600 2018).

In addition to reviewing the government regulation archives related to climate change, green development, and socioeconomic development plans issued at the Jiangsu provincial government and national levels, the Zhenjiang Development and Reform Commission also consulted opinions and suggestions from experts at universities and research institutes, as well as other related bureaucratic agencies in Zhenjiang. An extended consultation was held with the development and reform bureaus of twelve cities in Jiangsu Province. A total of thirteen comments and opinions—on subjects ranging from clarifying the definition of low carbon to adding extra indicators, such as the percentage of green areas in urban areas and the percentage of natural gas consumption in the local energy mix—were adopted before the proposed evaluation system reached the final review stage. The final version was submitted to the Jiangsu Provincial Quality and Technical Supervision Commission, which released it as a provincial standard (Zhenjiang Development and Reform Commission 2018).

This local initiative was successful because the low-carbon issue was important for Zhenjiang's image building through its engagement in the low-carbon pilot program, and because the Zhenjiang Development and Reform Commission prioritized the issue on its agenda (Interview 041826 2018). The bridge leader of the city's Development and Reform Commission's involvement in coordinating the implementation process further inspired local officials to explore new low-carbon initiatives and helped mobilize implementing resources, particularly implementation alliances. Zhenjiang's reputation for low-carbon development continues to motivate local bureaucrats and leaders to maintain the momentum.

Building energy efficiency and green building standards

Well before Zhenjiang's engagement in the low-carbon pilot city program, its Housing and Urban–Rural Development Bureau developed energy efficiency improvement programs to promote decarbonization. These initiatives also laid the groundwork for the city to build a cross-divisional alliance for its low-carbon experimentation. In 2011, the Jiangsu Provincial Housing and

Urban–Rural Development Commission selected Zhenjiang as a provincial demonstration pilot for building energy efficiency and green buildings in its central business district. In 2012, the Zhenjiang Leading Group for Building Energy Efficiency and Green Building, chaired by the vice mayor, was established.

Yet Zhenjiang was no closer to meeting its green goals until the Zhenjiang Housing and Urban–Rural Development Bureau took part in the Jiangsu Green Building Demonstration City Project in 2014. Over the next three years, Zhenjiang completed seventy-seven green building demonstration projects, which involved a construction area of 5.47 million m^2 (Zhenjiang Housing and Urban–Rural Development Bureau 2018, 160). The city continued to strengthen its regulatory measures by developing the Zhenjiang Green Building Development Plan (2016–2020), which sought to add another 9.68 million m^2 of newly built green building areas, and introducing green building indicators into the requirements for local land lease approval and fixed asset investments (Zhenjiang Housing and Urban–Rural Development Bureau 2018, 34–47).

Zhenjiang was able to enhance its initiatives related to building energy efficiency and green buildings for two reasons. First, the Zhenjiang Housing and Urban–Rural Development Bureau secured considerable financial resources for policy implementation by joining forces with the Jiangsu Provincial Housing and Urban–Rural Development Commission on its provincial demonstration projects. In 2014, the provincial government allocated 50 million RMB (US$8.07 million) for Zhenjiang building energy efficiency and green building demonstration projects; the municipal government provided an additional 60 million RMB (US$9.68 million) in matching funds to the projects (Interview 030700 2018; Zhenjiang Housing and Urban–Rural Development Bureau 2018).[2]

The second boost to low-carbon policy initiatives was the creation of the leading group chaired by the vice mayor. In 2015, the Zhenjiang Housing and Urban–Rural Development Bureau began using the group to coordinate actions with relevant actors from the building sector, such as research institutes, construction companies, and quality inspection divisions in the policy implementation process (Zhenjiang Housing and Urban–Rural Development Bureau 2018). This bureaucratic-level coordination mechanism facilitated local engagement in building energy efficiency and green building activities by setting up a comprehensive policy system consisting of a specialized fund; standards for design, construction, and post-construction evaluation; pilot projects; and action plans.

The Zhenjiang Housing and Urban–Rural Development Bureau has become more engaged in low-carbon policy experiments through its involvement in Zhenjiang's Leading Group for Low-Carbon City Construction since 2012. The green building policy sector also advanced as more political attention and additional executive resources became accessible after the Housing and Urban–Rural Development Bureau worked closely with the Zhenjiang Development and Reform Commission to implement the low-carbon policy experiments. The Zhenjiang Development and Reform Commission is one of the city's most powerful bureaucratic agencies, and it is well-connected to other local government agencies. The Zhenjiang Development and Reform Commission facilitated the development of energy efficiency and green building initiatives in its role as a major cross-division coordinator in low-carbon pilot city policy implementation (Interview 030700 2018).

The Housing and Urban–Rural Development Bureau submitted a request to the Zhenjiang Development and Reform Commission that green building indicators should be integrated into the policy system. This idea of strengthening the enforcement of green building requirements within the entire building process—from project approval, planning review, land leasing, project construction to completion inspection—appeared in the Zhenjiang municipal government's Opinion towards Green Building Promotion, and was listed on the working agenda of the Zhenjiang Leading Group for Building Energy Efficiency and Green Building in 2013 (Zhenjiang Housing and Urban–Rural Development Bureau 2018, 28).

The cross-divisional challenges of incorporating the compulsory green building indicators into action plans and government requests for proposals and land leases were finally resolved in 2018 when the Housing and Urban–Rural Development Bureau asked the Zhenjiang Development and Reform Commission for assistance, and expressed its willingness to push low-carbon initiatives in the building sector for the low-carbon pilot program (led by the Zhenjiang Development and Reform Commission) in return (Interview 030700 2018). The green building indicators were thus integrated into the policy system, particularly within government administrative procedures.

Joint actions were built among bureaucratic agencies to strengthen green building policy enforcement, which local developers and enterprises had challenged in the past (Interview 030700 2018). Although the Zhenjiang Housing and Urban–Rural Development Bureau had promoted local building energy efficiency and green building before the start of the low-carbon pilot city program, some of its most significant policy progress in recent years has come from its joint efforts with the Zhenjiang Development and Reform Commission. As of December 2017, Zhenjiang had 120 civil construction projects that met the green building standards, with a total construction size of 11.05 million m^2. During the implementation of the Zhenjiang Green Building Demonstration Project between 2014 and 2017, 6.19 million m^2 of new green building construction was initiated, and decarbonization measures, such as renewable energy utilization and energy-efficient building structures, were introduced into the construction process. These measures led to an annual reduction of 42,400 tons of carbon and 162,000 tons of coal equivalent (Zhenjiang Housing and Urban–Rural Development Bureau 2018, 166).

Creating a low-carbon development division

Since Zhenjiang launched its low-carbon pilot city program, almost all its new experimental initiatives have been mobilized using the same local leadership mechanism: top political leaders endorsed the issue area with their political support and delegated specialized policy tasks to local agencies inside and outside the government for implementation. After the mayor left office in 2016, the low-carbon initiatives continued, thanks to the efforts of a bridge leader at the Zhenjiang Development and Reform Commission. The bridge leader, with the help of the vice mayor, advanced local policy experimentation efforts, including setting up a specialized bureaucratic division for low-carbon development.

According to multiple sources from Beijing and Zhenjiang, unlike in many other local areas, the national bureaucratic reforms of 2017 and 2018—which shifted the Department of Climate Change from the National Development and Reform Commission to the Ministry of Ecology and Environment—did not bog down Zhenjiang's new low-carbon policy experiments (Interview 081724b 2018; Interview 231908 2018; Interview 041826 2018; Interview 082600 2018). During that time, Zhenjiang's policy experiments were extended to two new frontiers: the creation of a specialized low-carbon bureaucratic agency inside the government and pushing the local low-carbon city evaluation system to the provincial level.

Low-carbon energy transition matters formerly belonged to the Energy Conservation and Environmental Protection Division of the Zhenjiang Development and Reform Commission (Zhenjiang Institutional Organization Commission 2017). In December 2017, the Zhenjiang government replaced the Government Office's Ecological Coordination Division with a new Low-Carbon Development Division (located in the Zhenjiang Development and Reform Commission), which had four permanent staff members (one division head, one vice division head, and two other officials)

(Zhenjiang Institutional Organization Commission 2017), which gave Zhenjiang arguably the strongest bureaucratic-level executive capacity at that time among the four case cities.

The relatively strong executive capacity has enabled Zhenjiang to build connections with agencies at higher levels of government to create a wider low-carbon action network than many other local areas and regions pursuing low-carbon practices. Zhenjiang has connections with national-level agencies. As the central–local connection between the National Center for Climate Change Strategy and International Cooperation and the Zhenjiang government strengthened, through their strategic cooperation agreement on low-carbon development cooperation in 2016, Zhenjiang expanded its external connections and enhanced its image. For instance, the national center helped the Zhenjiang government organize meetings in Beijing to attract enterprises and institutions to attend the city's International Low-Carbon Technology Production Trade and Exhibitions between 2017 and 2019. At the exhibition, Zhenjiang signed nine strategic cooperation agreements with leading institutions and enterprises such as the United Nations Development Programme and Swiss ABB Group, with a focus on low-carbon energy, ecological protection, and low-carbon town construction. A total of sixty-three low-carbon industrial projects were signed at the exhibition, with a total investment of 55.9 billion RMB (US$9.02 billion) (Zhenjiang Development and Reform Commission 2017a).

Zhenjiang's changing political context in recent years has also created a concern: pursuing high-quality development by treating economic development as the central task on the local agenda. As the mayor noted, "high quality by no means is no development; slow development is not equal to high quality" (Zhang 2019). Since the Zhenjiang Bureau of Ecology and Environment took over climate issues from the Zhenjiang Development and Reform Commission in 2018, the city has yet to substantially decarbonize local industrial sectors. The local leadership of the Zhenjiang Party Committee and the Zhenjiang municipal government gave the green light to renewable manufacturing industries while explaining that it was impractical to set ambitious low-carbon targets (Zhenjiang Municipal Government 2020).[3] In 2021, Zhenjiang introduced the Ecological Civilization (Low-Carbon City) Construction Plan, which includes seventy-six low-carbon task forces such as those dedicated to renewable energy development, carbon peak action plans, and green building. While the city's Bureau of Ecology and Environment emphasized that the implementation outcomes of the task forces would be added to the bureaucratic agencies' annual evaluations, no ambitious carbon emissions reduction goals were proposed, only "meeting the targets assigned by the higher levels of government" (General Office of Zhenjiang Government 2021). It is also unclear whether the proposed low-carbon plans can be fully enforced or lead Zhenjiang to become truly decarbonized if these initiatives have not included the involvement of the power and industrial sectors, given their central impact on the city's total carbon emissions.

Conclusion

Despite its heavy reliance on coal consumption, Zhenjiang has been able to introduce institutional low-carbon policy initiatives. Like the other low-carbon pilots, such as Shenzhen and Nanchang, the Zhenjiang municipal government has created formal institutions that have sustained low-carbon policy engagement even after local political leaders left office. Much like its counterparts in Shenzhen, Zhenjiang has gone considerably further in its low-carbon policy engagement. Its approach demonstrates the potential for continued local policy engagement when a strong bureaucratic agency-based bridging leadership works with political leaders to introduce major new low-carbon institutions. The case further demonstrates local governments' capacity to push low-carbon city evaluation standards to higher levels of government.

Yet, it is hard to say how well the Zhenjiang Bureau of Ecology and Environment has built implementation alliances with the industrial sector since it took over climate issues from the Zhenjiang Development and Reform Commission in 2018. The Bureau of Ecology and Environment has been known to be not particularly strong at forming cross-divisional actions that have been shown to be crucial for low-carbon policy experimentation and implementation.

The Zhenjiang case also raises questions regarding its continued low-carbon engagement if key carbon emitters are not engaged in implementation alliances. The city has secured broad implementation alliances with major government and non-government actors across different levels of government, but not key industrial actors. Its energy development plan for 2021–2025, issued in 2022, excluded the power sector from its proposed coal consumption cap even though coal-powered plants account for over 75 percent of its total power capacity.[4] Given the substantial impact of carbon emissions from local power sectors, institutional efforts, such as creating a specialized government division or promoting low-carbon city evaluation standards, can risk becoming symbolic without sufficient industrial participation.

7
Sporadic Engagement
Xiamen

Although it is an environmentally friendly and economically advanced city, Xiamen has taken a cautious approach to engaging in low-carbon policy experimentation. Although the municipal government developed a plan to build a low-carbon city, its engagement in implementing low-carbon policies has been *sporadic*; few institutional efforts have been made to sustain its policy experiments. After the municipal government shifted the bureaucratic authority for overseeing low-carbon policy experimentation to a bureaucratic agency that showed no interest in promoting the issue, the city has struggled to maintain low-carbon policy engagement. This chapter describes Xiamen's engagement in low-carbon city policy experimentation from 2010 to 2018.

A Cautious Low-Carbon Practitioner

Xiamen, located opposite Taiwan, is known as an environmentally friendly coastal city in Fujian province in southeastern China. To uphold its environmental protection efforts, the municipal government requires environmental construction investment to be no less than 3.5 percent of the city's GDP (Pan and Zhuang 2018, 59). The city has also attracted international attention by hosting sustainable events such as the Mangrove Coastal Wetland Ecological Park Project and a 2020 zero-waste marathon (United Nations Environmental Programme 2019, 2020).

Like many other Chinese cities, Xiamen's local policy agenda has been heavily driven by rapid economic growth and urban expansion; the municipal government has not hesitated to grow the city's energy-intensive industries and promote real estate development. Xiamen Island, used mainly as a port and to bolster national defense, has been rapidly urbanized. Many coastal wetlands and bodies of water have been transformed into urban areas: its urban area almost tripled from 30.50 km^2 in 1987 to 85.10 km^2

Implementing a Low-Carbon Future. Weila Gong, Oxford University Press. © Oxford University Press (2025).
DOI: 10.1093/9780197757451.003.0007

in 2004 (Lin et al. 2013). The city is home to an array of energy-intensive industries producing chemicals, rubber and plastics, computers and other electronic equipment, and non-metallic mineral products: in 2015, these energy-intensive industries consumed the equivalent of over twenty-three million tons of coal and accounted for more than 74 percent of the city's total energy consumption (Pan and Zhuang 2018, 100).

Xiamen, with hundreds of kilometers of coastline and above-average sunshine and wind resources, is well positioned for decarbonization, but this potential has been largely untapped. The local transportation and industrial sectors relied primarily on oil consumption during the study period, and the city's economic growth has largely been linked to the consumption of fossil fuels (Dai et al. 2023). Oil products accounted for 36.94 percent of the city's total energy consumption and coal was the key energy source for the electricity consumed by the building sector, which accounted for nearly one-third of the city's carbon emissions in 2014 (Pan and Zhuang 2018, 121, 163). Although its per capita energy consumption decreased by almost 27 percent between 2010 to 2018, its total energy consumption increased by 66 percent during this time (see Figure 7.1).

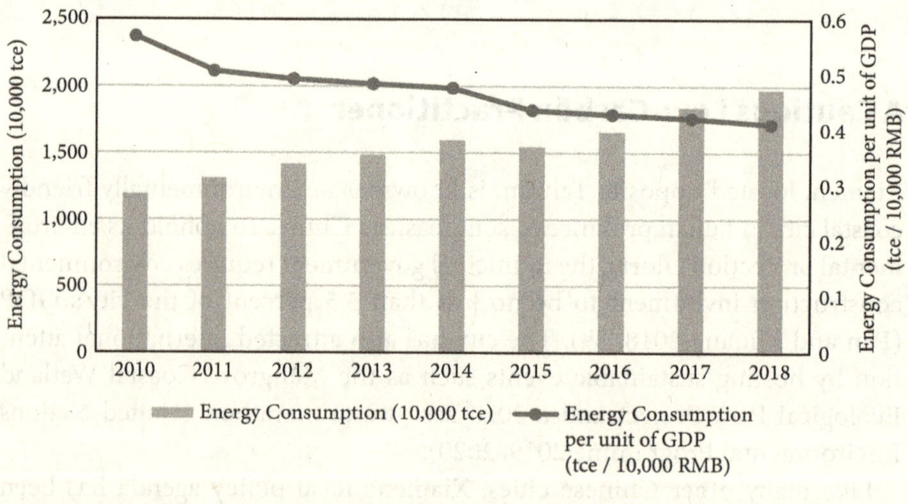

Figure 7.1 Xiamen's Energy Consumption and Energy Consumption per unit of GDP (2010–2018)

Note: Energy consumption per unit of GDP data for 2010–2015 is cited from the Xiamen Statistical Yearbook 2019. Data for 2016–2018 is estimated by the author based on the annual changes reported in the same source. Total energy consumption is calculated by the author using Xiamen's GDP data (2010–2018) and the corresponding energy consumption per unit of GDP, derived from both the Xiamen Statistical Yearbook 2019 and the author's estimates.

Source: Statistics Bureau of Xiamen Municipality and Survey Office of the National Bureau of Statistics in Xiamen (2019).

Xiamen introduced energy- and resource-intensive industries at the end of the 1980s when Haicang District, located in the west of Xiamen Island, opened as a Taiwanese Investment Business Zone. The district was designated an area of heavy industrial development for the petroleum and chemical industries in Xiamen's Overall Urban Planning in 1995 (Jiang 2009, 169). By 2006, Haicang District's tax revenue had grown year on year by 125.16 percent in real estate sales (Institute of Urban Environment Chinese Academy of Sciences 2007). But industrial companies' failure to properly enforce environmental regulations worsened the local living conditions. In 2007, Xianglu Petroleum Chemicals was found to be violating environmental standards and not using qualified environmental protection equipment; the company was emitting as much as ten times the normal level of pollution (Institute of Urban Environment Chinese Academy of Sciences 2007). The tension between polluted living environments, the real estate business, and heavy industrial development with lax environmental enforcement culminated in intense citizen opposition to building the Xiamen PX (para-xylene) chemical plant in 2007.

These protests encouraged the local government to reduce the economy's dependence on energy- and resource-intensive industries. In 2008, the Xiamen municipal government strengthened the environmental standards for local investment and construction project approval. It stipulated that it would not approve construction projects if they did not meet national industrial directives, required large amounts of energy and emitted high levels of pollution, did not meet clean production standards, or if local citizens strongly opposed them.[1] After it became a national low-carbon pilot city, the municipal government included low-carbon eco-city construction for the first time in Xiamen's Five-Year Plan for Economic and Social Development (Xiamen Municipal Government 2011). Yet it made cautious efforts to engage in low-carbon city policy experimentation in subsequent years.

The Leadership Path Toward Low-Carbon Policy Experimentation

On paper, Xiamen has done everything the central government has requested since it became a low-carbon pilot city in 2010. It has developed action plans, set goals, and created a leading group to coordinate and implement low-carbon initiatives. But in practice, it has been slow to translate these low-carbon goals into policy actions.

Bridging Leadership

Xiamen appears to have expressed initial interest in engaging in low-carbon development when it was selected as a national low-carbon pilot city. Yet without strong support from its government leadership, it did not display a desire to pursue low-carbon issues, and thus did not allocate executive resources to implement related policies. The leading group for low-carbon pilot city development in Xiamen was established but seldom met (Interview 191704 2018). Although it was chaired by local political leaders and had a municipal-level coordination mechanism for policy implementation, the group mainly functioned as a channel for circulating low-carbon-related documents within the government (Interview 191704 2018; Interview 120803 2018; Interview 102512x 2018). Given the lack of political support and capable bureaucratic leaders to mobilize the implementing resources needed to orchestrate policy implementation, Xiamen had insufficient engagement in low-carbon policy experimentation.

At first, the Xiamen Construction Bureau drafted the Xiamen Low-Carbon City Construction Plan as part of the city's application for the national low-carbon pilot city program. The Xiamen Jimei low-carbon demonstration program was one of the city's major proposed actions for its low-carbon policy experimentation; it involved incorporating low-carbon indicators into its land lease requirements (Xiamen Construction Bureau 2012). Yet none of the proposals were implemented, largely because the Xiamen Construction Bureau lacked the capacity to build implementation alliances for policy implementation (Interview 120803 2018; Interview 102512x 2018).

From the very beginning, Xiamen's low-carbon initiatives were challenged by a debate over whether the Xiamen Construction Bureau or the Xiamen Development and Reform Commission would take the lead in organizing and coordinating the policy experiment. Before the latter took over from the former, Vice Director Shuzhi Lin of the Xiamen Construction Bureau had been the bridge leader. The municipal political leader had made Lin responsible for the city's low-carbon pilot program (Interview 120803 2018). Lin was known for his expertise and leading role in promoting energy efficiency in buildings and construction (Interview 120803 2018; Interview 101708 2018). He brought together a small group of executive officials from the Xiamen Construction Bureau to develop a plan for Xiamen's low-carbon city development. Recognizing their limited executive authority to mobilize policy actions beyond the construction and building sector, the officials in the Xiamen Construction Bureau devised the plan with a focus on construction even though it set goals in a variety of issue areas.

Due to the lack of coordinating power to mobilize implementation alliances inside and outside the government, the Xiamen Construction Bureau was unable to implement the low-carbon standards it proposed for the Jimei low-carbon demonstration program. An informed local official from Xiamen explained that the bureau's proposed plan was bogged down because it challenged the interests of a related district that sought to pursue further development (Interview 101708 2018). Another official added that the low-carbon construction standards for the Jimei program were discarded because the Xiamen Construction Bureau was unable to enforce them against the local opposition (Interview 120803 2018).

The Xiamen Development and Reform Commission, which had a powerful coordinating role in macro-economic policy planning but showed no interest in low-carbon issues, took charge of the low-carbon policy experimentation in 2010. The Xiamen Construction Bureau merely reported progress in green building development and in building energy efficiency to the Xiamen Development and Reform Commission as required (Interview 120803 2018; Interview 191704 2018). After this shift in administrative authority, the Xiamen Construction Bureau focused on improving energy efficiency in the construction and building sector. In collaboration with the Xiamen Development and Reform Commission and the city's Economic Development Bureau, the Xiamen Construction Bureau promoted the Xiamen Green Building Action Implementation Plan in 2014 to ensure that new urban buildings comply with mandatory energy efficiency standards (Xiamen Municipal Government 2014). Building on this effort, green building standards were incorporated into the land lease requirements for all new government investments and civil construction projects. Xiamen was the first in Fujian Province to expand its green building standards from government-invested and large public buildings to civil buildings. Over 12 million m^2 of civil buildings passed the green building construction inspection during the study period, and over 200 million RMB (US$31.25 million) was allocated to reward companies that adhered to the green building standards (Xiamen Development and Reform Commission 2017).

The "Beautiful Xiamen" strategy: Decreased political attention to decarbonization

While the low-carbon issue emerged on the local agenda in the early 2010s, Xiamen's government was engaged in a leadership transition; the government leadership changed in 2013. As in other countries, local engagement in low-carbon issues can change significantly as key local government leaders shift

(Rabe 2004, 157). In China, the boundaries of mobilizing resources within local governments are not clearly defined; these resources are often reshuffled among bureaucratic divisions according to the political leaders' personal priorities (Personal Conversation 241326 2018; Interview 010712b 2018). The leadership transition in Xiamen had a huge bearing on its low-carbon policy experimentation, which the former leadership had endorsed but not yet implemented.

The new leadership sought to boost Xiamen's economic development. After Menghui Wang became Party secretary in 2013, local political attention turned to his favored policy program, "Beautiful Xiamen," which sought to double Xiamen's GDP per capita by 2021. Very little attention or implementing resources were devoted to low-carbon issues after that (Interview 030223a 2013; Interview 072526 2018; Interview 081724b 2018).

Wang established a leading group to implement Beautiful Xiamen within the Xiamen Party Committee and Xiamen municipal government, along with the corresponding implementation offices with specific task force teams in the municipal and township governments.[2] To attract investment and capital, the municipal government promoted large infrastructure projects to connect Xiamen Island to its mainland areas, and to connect the city with neighboring cities. In a 2013 speech, Wang asserted that "large infrastructure and major urban constructions are key to improving the city's carrying capacity and livability. Building a large number of rapid-transit systems, the areas inside and outside Xiamen Island can be better connected. All the bay areas in Xiamen should be better constructed to transform it from an island city into 'a bay regional city.'"[3] Over 100 billion RMB (US$15.6 billion) was invested in transportation construction between 2011 and 2015, which was 1.4 times the amount invested from 2006 to 2010 (Pan and Zhuang 2018, 131). Over the next few years, the Xiamen government invested more than 60 billion RMB (US$9.37 billion) in infrastructure construction projects, but only 1.6 million RMB (US$250,000) in low-carbon projects (Pan and Zhuang 2018, 57; Xiamen Academy of Urban Planning and Design 2017).[4]

Resource Mobilization

Xiamen's capacity to advance its low-carbon pilot city project has been constrained by limited access to implementing resources. Due to the lack of bridging leadership and political support, scarce resources, such as trained personnel and implementation alliances, cannot be easily mobilized. As of

2018, the city's low-carbon policy experiment was primarily led by people working temporarily for the government; specialists were seconded from a government-affiliated research institute.

While the Xiamen Development and Reform Commission acquired trained personnel, executive staff members lacked the authority to mobilize the implementation alliances that were crucial to sustaining low-carbon policy engagement. According to a source familiar with the implementation process, there was no external support to assist staff who sought to follow through with low-carbon policy initiatives (Interview 191704 2018). The source further commented that when organizing big low-carbon events in other cities, like those organized in Shenzhen, other local actors would normally serve as co-organizers but that was not the case in Xiamen (Interview 191704 2018).

While other pilot cities kept in close touch with the National Development and Reform Commission about the progress of their policy experimentation, Xiamen lagged in reporting its policy progress. It has maintained only a loose connection with the national level due to the lack of political attention and the frequent change in the bureaucratic officials responsible for the low-carbon pilot city program.

A source from Beijing interviewed for this study mentioned that Xiamen provided few updates to the national government during its low-carbon policy experimentation (Personal Conversation 081724a 2013). This source suggested I speak to the local official responsible for administering the city's pilot program, but the administrator declined my interview request, saying that no employees of the Xiamen Development and Reform Commission knew much about low-carbon issues. He had recently worked at the bureau and was not familiar with the issue either. Xiamen's administrator changed four times between 2010 and 2018 (Interview 191704 2018). Thus, the relevant officials were insufficiently informed to report on local progress to Beijing (Interview 081724b 2018; Interview 072526 2018; Interview 171019 2018; Interview 191704 2018). National coordination meetings for the pilot cities gave municipal governments the chance to build a direct connection with the center, which helped them obtain recognition for their progress and gauge the potential level of implementing resources from the central government including projects and financial support (Interview 081200b 2018; Song et al. 2020). The absence of key government leaders' participation in national-level coordination meetings indicated Xiamen's lack of political support and implementation capacity to pursue low-carbon policy experiments.

A new form of alliance?

Despite the absence of political support and agency-based bridging leadership within Xiamen's municipal government, new interest from the local business sector has engaged the local government in a new low-carbon initiative. In 2017, a local enterprise, Optikom, worked with the Xiamen Development and Reform Commission to develop the Xiamen Carbon Emission Management Platform. According to sources from Xiamen, Optikom initially proposed this new low-carbon program to a bureaucratic leader in the Xiamen Development and Reform Commission (Interview 081200b 2018). The Commission did not initially back the proposal, but after the company's chief executive officer managed to convince the Commission's vice director, it received bureaucratic-level support (Interview 081221 2018). After the Xiamen Government Office earmarked financial support for carbon monitoring reporting and verification in 2017 (General Office of Xiamen Municipal People's Government 2017), the Xiamen Development and Reform Commission, together with the company, introduced a Carbon Emissions Monitoring Platform involving over sixty energy-intensity enterprises the following year (each with an annual energy consumption over 5,000 tonnes of coal equivalent).

With the guidence of the Xiamen Development and Reform Commission, this program was able to navigate the bureaucratic process and secured 4.12 million RMB (US$643,750) in government funding to support software development, project management, and user training (Xiamen Development Research Center 2018). The Xiamen Development and Reform Commission organized three mobilization meetings, and all the participating enterprises received user training (Xiamen Development Research Center 2018). All data is directed to the government system to prevent leaks of enterprise data (Xiamen Development Research Center 2018, 3–5). Optikom initially intended to approach the Xiamen Development and Reform Commission about continued collaboration after the end of the initial one-year government funding to maintain and further develop the system, but long-term collaboration was not put in place (Interview 181200b 2018). As of 2018, the platform had collected carbon emissions data from sixty-eight energy-intensive industrial enterprises (Xiamen Development Research Center 2018, 1). In 2019, this carbon emissions management platform was selected as an industrial internet demonstration project and was awarded approximately 2 million RMB (US$312,550) for further development by the Fujian Industry and Information Technology Department (Fujian Industry and Information Technology Department 2019).

Conclusion

Although it has been receptive to environmental protection, Xiamen has pursued a cautious approach to its low-carbon policy experimentation. After being selected as a national low-carbon pilot city, it created its low-carbon city construction plan, formed a leading group for the policy experiment, and proposed a specific urban construction project to integrate low-carbon standards. However, many of its proposed action plans were not fully enforced, largely due to the absence of bridge leaders, particularly after changes in the political leadership and the shift of the bureaucratic authority overseeing low-carbon issues. Overall, Xiamen has only sporadically engaged in low-carbon initiatives and has created few institutions to support continued low-carbon policy actions. Xiamen has created forty-two low-carbon community demonstration projects in industrial parks and tourist venues since 2021 and was preparing to issue low-carbon evaluation standards for those projects as of 2023.[5] If Xiamen's government creates specialized low-carbon agencies to integrate these standards into its policy system and enforce them, this will signal it is moving away from sporadic engagement in decarbonization.

8

Rethinking Subnational Climate Leadership in China and Beyond

Subnational engagement in climate action has increased globally in recent decades; subnational leadership is now an essential component of climate governance (Betsill and Bulkeley 2006; Hsu et al. 2017; Hughes 2019; Jörgensen et al. 2015; Kelsey 2021; Koehn 2015). Recent research indicates that over 3,000 local areas and regions around the world have pledged to reach net zero, but fewer than half of them have set trackable targets that are essential for policy implementation (Song et al. 2024). Why have some local areas and regions been more successful than others at initiating and implementing climate policy actions?

China has recently emerged as a climate policy leader, particularly through its distinctive mode of "experimentation under hierarchy" (Bulkeley 2023; Heilmann 2008b). In response to Beijing's 2030 carbon peak and 2060 neutrality pledges, leading national bureaucratic agencies—the Ministry of Ecology and Environment and the National Development and Reform Commission—have launched multiple new experimental climate policy programs, similar to the one examined in this book, to encourage subnational engagement in addressing climate change. In 2022, the Ministry of Ecology and Environment, together with eight other national government agencies, introduced climate finance pilots in twelve cities and eleven city districts (Ministry of Ecology and Environment et al. 2022). In 2023, the National Development and Reform Commission introduced carbon peaking pilots in twenty-five cities and ten industrial parks (National Development and Reform Commission 2023a) as part of the national plan to select one hundred carbon peaking pilots in cities and industrial parks to explore local paths to achieving the carbon goals (State Council 2021).

In many respects, subnational governments in China are unlikely candidates for becoming climate leaders. They have faced common challenges, shared by subnational governments elsewhere, including insufficient political, fiscal, and bureaucratic resources to implement the proposed actions (Rabe 2004, 155–157). International observers often assume that Chinese

Implementing a Low-Carbon Future. Weila Gong, Oxford University Press. © Oxford University Press (2025).
DOI: 10.1093/9780197757451.003.0008

subnational governments simply implement central government orders and have limited political authority and fiscal resources to initiate policy changes. The fragmented government structure and corresponding bureaucratic competition in the Chinese system have created obstacles to cross-divisional collaborations within the government that are crucial for local policy implementation (Mertha 2008, 8). Nevertheless, some municipal governments have overcome the constraints of insufficient resources and authority to translate climate pledges into policy actions.

This book explores the role of subnational government leadership in climate policy implementation in China's centralized political system. I focus on local mid-level bureaucrats (bridge leaders) and their mobilization strategies, which are essential in initiating and sustaining low-carbon policy experimentation. The book makes three contributions that are pertinent to academics, policy makers, and policy practitioners. First, I examine the cases of four Chinese cities to unpack subnational climate policymaking and how each case translated low-carbon ambitions into actions. While political leaders, such as governors and mayors, are often in the spotlight and known to be important players in fulfilling the climate pledges, local bureaucrats undertake much of the work in implementing policy experiments but have received insufficient attention in prior work in this area.

I demonstrate that mid-level bureaucrats are instrumental, even more than political leaders, in orchestrating the implementation process that translates low-carbon pledges into policy action. Given their longevity in their localities, those local bureaucrats' insights into the local politics, policy, and policy process—as well as their connections inside and outside the government—have enabled them to mobilize the political, financial, and human resources needed to implement policies. My central argument is that the availability of bureaucratic entrepreneurs and their ability to mobilize resources condition local low-carbon policy implementation outcomes.

In a second contribution, I consider the strategies that bridge leaders have employed to mobilize implementing resources for low-carbon policy actions despite the constraints of limited political, financial, and human resources within local governments. To promote a new issue like low-carbon policy, local governments need to tackle three key challenges in the policy process: (1) overcoming the apathy toward the new issue in the bureaucratic system, (2) providing feasible solutions to policy problems, and (3) winning over the potential opposition in the policy process. I find that local bridge leaders use three strategies to mobilize these resources: obtaining support from top political leaders, creating a group of trained personnel who are qualified to provide and implement feasible policy solutions, and building implementation alliances to overcome the challenges as they explore local paths to decarbonization.

Third, I examine how (and why) some local areas are more successful than others at initiating *and sustaining* low-carbon policy experimentation. While low-carbon policy experiments are known to be hard to sustain and vulnerable to changes in political leaders, I find that local policy experiments can be sustained as long as bridge leaders have built a degree of policy institutionalization consisting of standards, regulations, and legislation—particularly when paired with a group of trained personnel for enforcement—into the implementation process.

While political leaders are often unmotivated to establish new formal institutions out of policy experiments (Fewsmith 2013, 6), local bureaucrats tend to be more inspired to maintain and institutionalize such experiments. Doing so allows them to defend and validate their agencies' previous efforts to promote low-carbon initiatives. In this book, I demonstrate how bureaucratic entrepreneurs create new climate institutions and facilitate low-carbon policy institutionalization through experimentation. Through these efforts, they generate incremental but groundbreaking changes in China's climate governance.

This chapter reflects on the book's findings and rethinks subnational climate leadership in China and beyond. It explores three questions: What does my study tell us about the role of local bureaucrats in China's climate policymaking more generally? What is the future of subnational climate leadership in China? What are the implications for sustaining climate action in China and globally amid rising geopolitical tensions?

Subnational Climate Leadership: Taking Bureaucrats Seriously

There is considerable debate over whether a centralized or decentralized approach to decarbonization is more effective in China. Past work on policy experimentation in the country often concludes that local engagement in experiments is closely linked to pressure from the central government or is a result of their patronage ties with top central political leaders (Ding 2022; Habich-Sobiegalla 2018; Heffer and Schubert 2023; Kostka et al. 2019; Ran 2013; van der Kamp 2023; Wang 2013). While recognizing the national government's role in providing guidance and connecting subnational governments to international actors in the early stage of the policy process, this research finds that the outcomes of the low-carbon pilot city programs examined in this book are mainly conditioned by *local government leadership*. They are less the result of central government pressure and more heavily influenced by the presence of local bridge leaders in the government and their capacity to mobilize resources to initiate and maintain policy actions.

Bureaucrats have long played an instrumental role in China's international climate engagement. At the initial stage of its international climate engagement starting in the 1980s, bureaucrats—mainly from the Ministry of Foreign Affairs, the State Science and Technology Commission, the State Meteorological Administration, and the State Council's Environmental Protection Commission—influenced the country's position in climate negotiations (Economy 1997). Their engagement with the West enabled China to access international technical and financial support for development (Economy 1997). Bureaucrats at the subnational level are playing increasingly important roles in local policy processes, which include responding to central policy directives, formulating local policy incentives, facilitating institutional changes with allied key local actors, and orchestrating policy implementation (Ma 2017; Yasuda 2015; Zhang and He 2024). Studying local bureaucrats' underexplored bridging role in enacting these implementation processes advances our understanding of the dynamics of subnational climate policymaking and divergent local paths toward decarbonization.

My theoretical approach to understanding uneven subnational engagement in low-carbon policy experimentation focuses on local government leadership. I argue that bureaucratic entrepreneurs, as bridge leaders, are capable of pursuing various strategies to promote institution building as they work to implement durable policy actions. The *Bridging Leadership* mechanism helps introduce and sustain policy engagement that is crucial for long-term policy change. The rest of this section discusses how bridge leaders emerge, what motivates them, and the conditions that facilitate or hinder their success.

How bridge leaders emerge and thrive

Bridging Leadership needs the right conditions to develop and thrive. The emergence of a bridge leader can be compared to planting a tree: while a seed is germinating, external conditions such as the temperature, moisture levels, and location can foster or impede its growth. While there is a degree of unpredictability regarding the specific person or area in which bridge leaders may emerge, none of the relevant factors is random: who are the best candidates for bridge leaders in the government, what motivates them to become bridge leaders, where are they most likely to emerge (at what level of government and in which bureaucratic agencies), and under what conditions are they more likely to thrive.

Similar to Western countries such as the United States and Israel (Frisch-Aviram et al. 2018; Rabe 2004, 207), subnational governments in China, particularly at the city level, are more favorable environments than the central government for bridge leaders to emerge and promote new issue areas such as climate change. Unlike local political leaders who rotate every few years or street-level bureaucrats who lack decision-making authority, mid-level local bureaucrats are best placed to become bridge leaders. Even though mid-level bureaucratic leaders do not necessarily occupy high-profile political positions such as mayors, over time they gain decision-making power that enables them to command human and financial resources within their bureaucratic agencies that are crucial for mobilizing policy action in a new issue area.

A variety of factors can motivate bridge leaders, including their exposure and expertise in the issue area, their need to gain recognition from their superiors or higher levels of government for their personal or agencies' achievements, as well as potential self-serving political and economic gains (Lewis et al. 2022; Teets 2015). For new issues like climate change, bridge leaders often arise in bureaucratic units that expose local officials to low-carbon ideas through related work experience, training, or participation in international programs. This early exposure enables them to acquire new knowledge of or expertise in low-carbon issues that they can frame as policy solutions and employ to address local challenges. In this research, I find that local bureaucratic entrepreneurs supporting low-carbon development are often officials who have been involved in macro-economic policy planning or energy efficiency initiatives and participated in climate training programs provided by the national government and its international partners.

Given the smaller size of subnational governments, such as those at the city level, local bureaucrats also tend to work together more, which can encourage them to be more cooperative in the policy process. For example, subnational government agencies in China, such as local development and reform commissions, have fewer employees than their national counterparts, and the bureaucratic layers between the bureaucratic and political leaders are much thinner. Their deep experience in the bureaucratic system gives local bureaucratic entrepreneurs insights into political opportunities and policy connections when strategizing their policy options (Kingdon 1995, 33).

The bureaucratic positions that bridge leaders occupy usually help them identify opportunities to introduce new policy ideas to their superiors or higher-level governments as they translate such ideas into action. Bureaucratic entrepreneurs are often skilled at navigating bureaucratic networks to mobilize the necessary political support for policy change (He 2018; Zhu and Xiao 2014). Trusted mid-level bureaucrats often serve as policy advisors to local political leaders or their superiors on new issue areas in which they have invested time to learn about. These bureaucratic leaders can secure political support by providing policy advice to political leaders and reaching out to other key actors who have the political, financial, and legislative resources to build implementation alliances. When bridge leaders occupy strategic and coordinating bureaucratic positions, such as development and reform commissions in China, they tend to be more likely to form intra-governmental alliances for policy implementation.

When bridge leaders can obtain political support from and forge alliances with (and among) key actors, local engagement in low-carbon policy implementation is stronger (see Table 8.1). This suggests that even in China's centralized political system, strong local bureaucratic entrepreneurship facilitates policy implementation.

Many structural factors can undermine bridge leaders' success, particularly if the domestic political environment restricts the growth of entrepreneurship within the bureaucratic system. A risk-averse work environment tends to make local bureaucrats less motivated to think outside the box; policy experiments always entail political risks since local bureaucrats can easily be blamed for failed policies and approaches (Ang 2016, 248–249). It can also be more challenging for potential bridge leaders to build alliances within the government; this often involves persuading their superiors and counterparts in other bureaucratic agencies (who may be more risk averse) to engage. Prior research notes that Beijing's anti-corruption campaign has enhanced environmental policy enforcement but disincentivized local bureaucrats from taking risks in policy experimentation (van der Kamp 2023, 118). Local bureaucrats involved in policy initiatives are found to be less useful at providing innovative policy solutions "under pressure" (Heffer and Schubert 2023), as there is a strong temptation to engage in actions that give the *appearance* of adhering to directives from Beijing.

Therefore, bridge leaders are often those who have had early exposure to climate change issues and hold decision-making positions in subnational governments—particularly within agencies with cross-divisional coordinating functions—that enable them to explore and try out new solutions to local policy challenges.

Table 8.1 Comparative Analysis of Local Bridging Leadership and Resource Mobilization in the Implementation of Low-Carbon Pilot City Programs across the Case Cities (2010–2018)

Pilot City	Bridge Leader	Political Support	Implementing Resources		Policy Engagement
		Political endorsement from local political leaders	Number of trained government personnel implementing the low-carbon pilot city program	Key implementation alliances	Standards, Regulations, Legislation
Shenzhen	Vice Mayor	Yes	4	Shenzhen Development and Reform Commission, Shenzhen SEZ Development and Construction Group, Shenzhen Municipal People's Congress, Shenzhen Building Research Institute, local universities, and National Development and Reform Commission	High
Nanchang	Chief Economist	Partial	2	Nanchang Development and Reform Commission, Nanchang Finance Bureau, Nanchang Municipal People's Congress, Jiangxi Provincial People's Congress, Nanchang Carbon Development Promotion Association, and National Development and Reform Commission	Medium-Low
Xiamen	None	Partial	1	Lack of implementation alliances (government–business alliance was temporary)	Low
Zhenjiang	Deputy Director of the Development and Reform Commission	Yes	4	Zhenjiang Development and Reform Commission, Zhenjiang Transportation Bureau, Zhenjiang Housing and Urban–Rural Development Bureau, Zhenjiang Quality and Technical Supervision Bureau, Jiangsu Provincial Quality and Technical Supervision Commission, township-level governments, National Development and Reform Commission, and National Center for Climate Change Strategy and International Cooperation	Medium-High

Note: Xiamen had a bridge leader in the Xiamen Construction Bureau when the city applied to become a low-carbon pilot, but authority over low-carbon issues shifted to the Development and Reform Bureau soon after it was selected as a pilot. Nanchang lacked trained personnel to implement low-carbon policies after its bridge leader left the government in 2017.

The Future of China's Subnational Climate Leadership

Prior research on addressing climate change and environmental challenges, especially in urban areas, has highlighted several ways in which governments can transform the governing structure toward decarbonization, such as by challenging the underlying socioeconomic structure that maintains carbon-intensive capitalism, changing current social–material relations, fostering a group of trained specialists for decarbonization through experimentation, and contextualizing international learning into local solutions that cannot be resolved through international climate negotiations (Bulkeley 2023; Cashore et al. 2019; Dauvergne 2018; Hoffmann 2011; Hughes 2019; Paterson 2021; Sabel and Victor 2022, 164). Given the major institutional changes that these studies recommend governments undertake in climate experimentation, it is important to understand the conditions that shape subnational climate leadership's ability to make policy changes.

China's subnational climate policy initiatives have faced three main sets of challenges and opportunities since the 2018 climate institutional reform. First, subnational climate policy implementation has stagnated in tackling climate change and decarbonization. On the one hand, President Xi Jinping has reiterated that "[China's efforts] to deal with climate change [are] not imposed by others but promoted by ourselves," signaling the national government's commitment to achieving its climate pledges, at several international and domestic events.[1] On the other hand, removing the two major trackable low-carbon targets—energy intensity and carbon intensity—from their 14th Five-Year Plans (2021–2025) for Economic and Social Development signaled local governments' caution in tackling climate change, which makes policy progress more difficult due to a lack of measurable targets to guide implementation. Indeed, following these plans, local governments have largely rolled back their individual (sometimes ambitious) carbon peak pledges to comply with the national 2030 carbon peak timeline instead. In addition, although the national government issued standards for measuring carbon emissions, reporting, and verification in ten industrial sectors a decade ago (National Development and Reform Commission 2014), local governments have yet to create regulatory standards to audit and report greenhouse gas emissions from different sectors or to evaluate local carbon peak pathways (Wang 2020). Local bureaucrats tend to appease higher levels of government by engaging in performative or "face" initiatives when they lack the capacity to deliver the public goods as expected (Ding 2022; Teets 2015). Beijing's strong political pledges, coupled with weak policy incentives at the subnational level to achieve climate goals, can lead to local performative engagement: local officials echo central political rhetoric but do not deliver substantial policy outcomes.

Second, subnational climate leadership clearly faces several challenges in helping China achieve its climate goals. A key question is how best to institutionalize useful experimental policy practices into the existing policy system, particularly when bureaucratic competition is high in the upper levels of government. The frequent changes in bureaucratic authority over climate issues have also complicated subnational climate engagement. In 2018, the national government moved the Department of Climate Change from the National Development and Reform Commission to the Ministry of Ecology and Environment. Responsibility for climate change policy has been split between the National Development and Reform Commission and the Ministry of Ecology and Environment since 2021 when the National Development and Reform Commission was designated as the leading agency to coordinate carbon peaking initiatives nationwide (State Council 2021). At the time of writing, the National Development and Reform Commission oversees meeting the carbon pledges, while the Ministry of Ecology and Environment focuses on issues related to the national ETS and climate adaptation. The key issue will be how the two agencies coordinate to achieve the country's carbon goals and whether the latter gives its Department of Climate Change sufficient resources and authority to make decisions and facilitate low-carbon policy institutionalization.

Third, non-government actors must play a more substantial bridging role in enhancing climate governance to achieve decarbonization. Such actors have been increasingly engaged in China's environmental and climate governance, such as in setting forestry standards, environmental litigation, environmental information transparency, international climate negotiations, greening the supply chain, and international energy investments (Gong and Lewis 2023; Guttman et al. 2018; Liu et al. 2017; Stern 2013; Yang 2023). For instance, professional associations have recently emerged as leading figures in coordinating subnational climate initiatives. They have played an important role in monitoring the implementation of new government climate policies (Guttman et al. 2018); thus, strengthening their engagement may provide more flexibility for climate initiatives when government leadership is not available. For example, the Shenzhen Green Finance Committee was established to facilitate the implementation of the country's first local green finance legislative regulation, passed by the Shenzhen Municipal People's Congress in 2020, called the Regulations of Shenzhen Special Economic Zone on Green Finance (Shenzhen Municipal Administration for Financial Regulation 2021;

Shenzhen Municipal People's Congress 2020). This law required seventy-six local financial institutions—including banks, insurers, securities firms, fund firms, and trust firms—to disclose environmental information.[2] The director of China Development Bank's Shenzhen branch chairs the association, which includes over twenty other local banks; it has become a key actor in coordinating government and business green finance initiatives, such as offering the finance sector environmental risk evaluations and environmental information disclosure training. Therefore, non-governmental organizations have the potential to become a source of *Bridging Leadership* as they continue to strengthen their resource mobilization capacity, such as by facilitating enhanced government–business alliances on decarbonization.

Strengthening Subnational Climate Policy Implementation in China

Scholarly and public debates explore whether authoritarian political systems are more effective than their democratic counterparts at environmental governance (Eaton and Kostka 2014; Li and Shapiro 2020; Mertha 2009; Moore 2014).[3] Prior research indicates that centralized policymaking in China can be effective at producing policy output but not policy *implementation* (Gilley 2012). As this book suggests, the presence of bridge leaders—and their persistence in pushing low-carbon issues and proposed solutions—is central to the entire experimentation process from agenda setting to implementation. How can subnational governments in China be more motivated to enact and implement climate policies?

Change is needed on three fronts. First, insufficient human capacity has been a major cause of disjointed local implementation (Jahiel 1998); surmounting this challenge requires greater bureaucratic capacity and accountability. A bureaucratic division of climate change (low-carbon development) staffed by trained bureaucrats with posts (*bianzhi*) overseeing climate policy implementation should be created in every municipal government to improve the quality of city-level energy and emissions data. Most emissions data below the city level is currently self-reported. These inadequate energy statistics, together with insufficient sectoral and energy-type data, constitute a key challenge for climate policy implementation in cities (Liu et al. 2022). The barriers to sharing emissions data across bureaucratic agencies also need to be removed (Ministry of Ecology and Environment 2023b). There are too few local climate specialists (China Council for International Cooperation on Environment and Development 2019), which makes overseeing policy implementation particularly challenging and may undermine China's climate policymaking.

It is also in the national government's best interests to promote subnational engagement in the creation of more low-carbon institutions before 2030. Old bureaucratic organizations may not be best placed to advocate new and innovative tasks (Lieberthal and Oksenberg 1988, 414; Dubash et al. 2021) because individual leaders may struggle to overcome the deep-seated interests in their bureaucratic networks and relationships. Therefore, although many local areas and regions have made existing bureaucratic agencies responsible for climate issues, such as the environment and resources units under the ecology and environment bureaus, it would be more effective to create a specialized bureaucratic agency responsible for climate change and low-carbon policy implementation. Cities that have created new formal government posts on climate change are better able to fund low-carbon staffing and decarbonization in the long run. Those that can rely on full-time personnel working on climate policy are better positioned to increase their accountability and capacity to mobilize resources for enhanced climate policy implementation.

Yet institution building alone is insufficient to generate long-term change without specialized personnel working on implementation. Institutions need specialists to ensure policies are enforced, and resources need to be mobilized to sustain the momentum on decarbonization. For example, after the national government shifted administrative authority for climate change from the National Development and Reform Commission to the Ministry of Ecology and Environment, cities with no climate specialists stalled in their implementation of low-carbon actions. It is worth noting that it may take three to five years to train a new group of bureaucrats with the knowledge, experience, and implementation capacity to deal with carbon emissions reduction assessments and evaluations (China Council for International Cooperation on Environment and Development 2019).

Cities that already had a group of climate specialists, such as Shenzhen and Zhenjiang, were able to minimize the policy disruption caused by this institutional reform and maintain local engagement as new climate institutions were created and paired with trained officials to carry out low-carbon initiatives. The Zhenjiang government's approach could be replicated throughout the country: it created a dedicated bureaucratic division for low-carbon development, staffed with trained personnel appointed through rearranged established posts (*bianzhi*), to oversee climate change and decarbonization issues.

Moreover, bureaucratic agencies that have a super-coordinating function, such as development and reform commissions, are better able than implementation-oriented agencies to push issues onto the government policy agenda in China (Lieberthal and Oksenberg 1988, 414). My study shows that bureaucratic leaders in the latter type of agency may encounter challenges if the former is not part of the implementation alliance for low-carbon projects that need cross-divisional collaboration.

Second, subnational governments need more leverage to mobilize resources. Fiscal difficulties have been recognized as a key barrier preventing subnational governments from implementing proposed policy actions (Lo et al. 2019). Unlike low-carbon technology industries—such as the wind and solar sectors, which have been targeted with multiple policy incentives from both national and subnational governments to forge competitive advantages in globalization (Nahm 2017)—subnational climate initiatives have been underfunded. The only specialized climate funding that subnational governments can apply for is the China Clean Development Mechanism Fund, which supports a limited scope of subnational initiatives such as research, educational, and capacity-building activities (Ministry of Finance of People's Republic of China 2020).

A national funding scheme would place local bureaucrats in a better position to leverage fund matching from subnational governments for climate policy implementation. More channels for financial resources need to be introduced to foster subnational government engagement in decarbonization. Insufficient national government funding has constrained local officials' ability to promote low-carbon initiatives beyond the stage of research and educational programs (Interview 061900, 2018). Passing the funding responsibility onto subnational governments is likely to be counterproductive: missed incentives, combined with the weak bureaucratic standing of the national-level Department of Climate Change, make it harder for local bureaucrats to convince political leaders to place low-carbon issues on the government agenda. The lack of dedicated funding from the central government decreases bridge leaders' ability to mobilize political support or matching financial resources from their own governments. This can weaken China's ability to realize its national climate ambitions because funding shortfalls make it harder for subnational governments to maintain their engagement in climate policy implementation.

Third, the Chinese government needs to strengthen central-level climate coordination to select feasible policy solutions generated by local policy experiments to inform the process of low-carbon policy institutionalization nationwide. National policy reforms can be triggered from the bottom up if national support and local needs can be coordinated to scale up the policy lessons learned from experimentation nationwide (Teets and Hurst 2015). Even though Beijing has tended to coordinate policy experimentation to seek feasible solutions to inform national policymaking, local policy practices may not be elevated to the regional or national levels. Policy experiments can also be quietly discarded without scaling them up nationwide (Heilmann 2008b). Some experiments have failed due to insufficient central coordination to ensure local discretion to sustain experimentation or resolve subnational disputes over standardization (Yasuda 2024). Policy practices that have been tested in one locality may not be applicable elsewhere due to different socioeconomic conditions or regional competition; therefore, central coordination is critical.

While Chinese cities have demonstrated their ability to experiment with new decarbonization policies, the diffusion of local experimental practices across regions or to higher levels of government has been limited. The ultimate goal of low-carbon policy experimentation is to integrate the policy innovations generated in the policy experiments into the conventional governing system (Hoffmann 2011, 154). According to data from the Ministry of Ecology and Environment (2023b), after nearly ten years of low-carbon policy experimentation, various local experimental initiatives—including much-needed regulatory measures such as greenhouse gas emissions accounting and a management and target responsibility system for controlling greenhouse gas emissions—are available but need further national coordination to streamline the respective standards and institutionalize them into the policy system.

If local leaders want to introduce policy changes in a political system with rigid path dependence, they must identify ways to integrate their initiatives as feasible policy solutions to address emerging challenges in the system. For example, the "1+N" climate policy system, which the national government pledged to build to coordinate decarbonization efforts in various policy sectors, could help bridge leaders determine which institutions they want to work with to integrate the solutions identified in policy experiments for addressing emerging problems in the policy system.[4] Similar lessons can be learned from Zhenjiang as it scaled up its low-carbon city standards to the provincial level. The long-term effects of such experimental policy achievements depend on the extent to which the national government can work with local bureaucrats to identify successful policy solutions and coordinate low-carbon policy institutionalization by using lessons learned from the experiments to inform the national climate policymaking process.

The central government has encouraged lower-level governments to engage in low-carbon legislative efforts (Tian 2018), but this has been untenable. It is also unclear how subnational low-carbon legislative efforts can inform the national climate legislative process. While it was important that a process was created to coordinate climate change initiatives, the national Department of Climate Change's relatively weak standing in the bureaucratic system has inevitably slowed the national climate legislative progress. Such limitations can be overcome by installing a national cross-division coordinating mechanism chaired by top political leaders in Beijing—similar to the National Leading Group for Carbon Peak and Carbon Neutrality—to form horizontal alliances between departments such as the National Development and Reform Commission and the Ministry of Ecology and Environment as well as the National People's Congress to advance the national legislative process of addressing climate change.[5]

Future national government efforts must also focus on creating further policy incentives to boost low-carbon engagement from both government and non-government actors across the board. For instance, Climate Envoy Xie Zhenhua noted in 2021 that the central government planned to create mid-term targets and roadmaps for carbon peak goals.[6] Political leaders and bureaucrats in provincial and municipal governments need low-carbon training to familiarize themselves with the climate change issue and the policy incentives associated with identifying opportunities to mobilize implementing resources. Bureaucratic evaluations should also be updated to reflect rewards and penalties related to their involvement in decarbonization efforts. Focusing on reforming these areas, in combination with expanding implementation alliances inside and outside the government, will help foster bridging leadership and subnational climate engagement to realize China's carbon peak and neutrality goals.

Sustaining Subnational Climate Engagement Beyond China

Although this book focuses on China's domestic climate policies, its political and policy implications apply more broadly. China's subnational climate engagement at the international level is increasing. Government officials below the national level are participating in bilateral collaborations to tackle climate challenges. The international implications of subnational climate engagement are significant because national targets are meaningless without effective subnational implementation.

Subnational policy entrepreneurship in climate actions

Countries and regions across the world have also pledged to transition toward a carbon-neutral world, positioning subnational governments as crucial for policy implementation. In recent decades, a substantial body of literature has examined how policy entrepreneurship has facilitated policy responses, including to climate change (Anderton and Setzer 2018; Frisch-Aviram et al. 2018; Goyal et al. 2020; Green 2017; Hultman et al. 2020; Huitema et al. 2018; Mintrom and Norman 2009; Zhao et al. 2016). Scholarly research has thoroughly documented subnational policy entrepreneurship in climate policy beyond China. In Europe, city networks have adopted a form of policy entrepreneurship to mobilize subnational climate initiatives from across different levels of government. The climate initiatives spurred by different networks have also led to divergent engagement between pioneering cities and smaller, less ambitious ones, which often lack the capacity and commitment to effectively implement climate goals (Kern 2018). In the United States, non-state actors such as states, cities, and businesses have pledged to continue their involvement in addressing climate change in response to President Trump's decision to withdraw the country from the Paris Agreement in 2017.[7] Despite continued subnational actor support for climate actions, regional disparity in climate engagement as well as symbolic subnational climate engagement have been acknowledged as obstacles to implementing the United States's national climate strategy (Rabe 2004, 155–157; Hopper and Swift 2021). Government agency-based entrepreneurship, however, has been identified as a major driving force behind California's climate leadership (Meckling and Nahm 2021). In Canada, research shows that policy entrepreneurs within political parties play a crucial role in sustaining decarbonization initiatives at the subnational level—particularly by framing policy issues as beneficial to key stakeholders for coalition building (Millar 2020). Thus, while local governments have primarily driven the policy entrepreneurship developed in the Chinese cities examined in this book, subnational policy entrepreneurship for climate action outside China has been led by a more diverse set of actors—both within and outside government.

Several interesting questions emerge for future research on subnational policy entrepreneurship for climate change. Under what conditions can subnational actors help build climate policy institutionalization? Who are the implementation bridges in climate policymaking outside China? What governing structure would best address the uneven levels of policy engagement between leading and lagging regions? How can subnational climate engagement inform higher levels of government in ways that strengthen multi-level climate governance as a whole?

This study highlights that the challenge in China is not simply whether Beijing can get subnational governments on board to decarbonize, or whether successful subnational climate initiatives will spill over to other areas. The difficulties associated with institutionalizing useful local practices and scaling them up to the national level discussed above also reflect a larger question facing China's subnational climate engagement—the extent to which it can influence national climate policy and strategy, particularly given the increasing concerns about geopolitics and climate change. It is unclear how much Beijing will allow subnational governments to engage in climate diplomacy, given that national-level bureaucratic agencies are assumed to dominate international climate negotiations.

Subnational climate leadership in an era of increasing geopolitical tension

National and international politics touch on the limits of subnational authority in pursuing climate leadership both domestically and internationally. First, rocky international relations (particularly between China and the United States) have affected subnational climate engagement at the international level. The deteriorating bilateral relations during the first Trump administration almost entirely stalled the bilateral subnational climate collaborations proposed during the Obama administration. The launch of the climate-smart/low-carbon city initiative in 2015, which was proposed and written into the U.S.–China Joint Announcement on Climate Change in 2014, motivated Chinese and American provinces/states, cities, districts, and counties to propose joint efforts to maintain subnational climate ambitions, share best practices, seek solutions to reduce carbon emissions and enhance climate resilience, and facilitate bilateral climate collaboration. Following the China–U.S. Climate Leaders Declarations signed in 2015 and 2016, a dozen Chinese cities created the Chinese Alliance of Peaking Pioneer Cities in 2016 and pledged to achieve their carbon peaks before or by the national target year of 2030. During the first Trump administration, no updates could be found on the alliance's development or proposed bilateral subnational climate collaboration. The United States rejoined the Paris Agreement in 2021 under President Biden; the U.S.–China joint declaration in Glasgow that year briefly mentioned enhancing local governments' climate participation but did not lead to concrete results.

Soon after U.S. House Speaker Nancy Pelosi's visit to Taiwan in 2022, Beijing suspended military communications with the United States, as well as bilateral cooperation on global governance issues including the climate, even though climate change is one of the few areas in which the two countries can remain engaged despite the tensions in their relationship. There were no signs of resumed bilateral subnational climate engagement until California Governor Gavin Newsom's visit to China in 2023. His visit prioritized California–China climate cooperation and yielded several signed Memoranda of Understanding with Chinese national-level bureaucratic agencies and subnational governments on a variety of issues including decarbonization in the power and industrial sectors, green shipping, and the carbon market.[8] This visit also paved the way for the Sunnyland Statement on Enhancing Cooperation to Address the Climate Crisis, which emphasized subnational cooperation as one of eight key issue areas on which the two countries agreed to collaborate (U.S. Department of State 2023). Implementing this agreement will not be possible without the leadership of pro-climate subnational governments and their sustained international engagement on climate change.

Second, domestic politics also matters. Under the second Trump administration, leading pro-environment states such as California, New York, and Vermont have faced challenges from Washington, DC. The federal government is questioning their authority to enforce stricter environmental regulations on local business sectors for addressing climate change, including measures based on the polluter-pays principle and the cap-and-trade scheme (White House 2025). In the past, U.S. state governments' direct work with other national governments has also been challenged on the basis that it violates constitutional restrictions on the authority of state governments encroaching on federal authority regarding international treaties and alliances (Rabe 2004, 165). Therefore, politics at the national and international levels can shape subnational governments' ability to pursue their climate leadership both domestically and internationally.

Future research could examine how linkages between the local and international levels can influence local government responses to climate change. It could also explore how much leeway subnational governments have in China and the United States with regard to climate diplomacy, using the implementation of the proposed subnational bilateral efforts as case studies. Such work will help establish whether (and how) the United States and China continue their strategic collaboration, particularly on common challenges that cross borders such as climate change, despite competition (Campbell and Sullivan 2019; Jaros and Newland 2024; Karplus et al. 2021; Schreurs 2024), and what role subnational governments can play in linking climate actions domestically and internationally amid intensifying geopolitical tensions.

The world is monitoring China's progress on its climate pledges. The extent to which it can achieve its low-carbon goals relies not only on Beijing's commitment but also on how it can empower subnational climate leadership in policy implementation, especially by implementing the conditions to nurture the thriving bridge leaders described in this book. I articulate the lessons learned from the low-carbon pilot city experimentation in China, emphasizing the role of mid-level bureaucrats within municipal governments as well as their mobilization strategies. Those who seek to understand why local implementation succeeds or fails should examine *Bridging Leadership*—and invest in these leaders to enhance climate policy implementation moving forward.

APPENDIX Distribution of Interviews

Sources	Numbers
Provincial Governments	10
Municipal Governments	28
Township Governments	4
Think Tanks	23
Universities	19
Enterprises	20
Non-Governmental Organizations	8
International Organizations	6
Media	4
Others	5
Total	**127**

Note: Government-affiliated research institutes at the central and local government levels are counted as think tanks.

List of Cited Interviews and Personal Conversations

Interview 010712a. 2013. Shenzhen.
Interview 010712b. 2018. Shenzhen.
Interview 020600. 2018. Zhenjiang.
Interview 02232426x. 2018. Nanchang.
Interview 02232426y. 2018. Nanchang.
Interview 030223a. 2013. Xiamen.
Interview 030700. 2018. Zhenjiang.
Interview 041826. 2018. Zhenjiang.
Interview 061900. 2018. Nanchang.
Interview 072526. 2018. Beijing.
Interview 081200a. 2013. Nanchang.
Interview 081200b. 2018. Nanchang.
Interview 081221. 2018. Xiamen.
Interview 081724b. 2018. Beijing.
Interview 082600. 2018. Zhenjiang.
Interview 100824. 2018. Nanchang.
Interview 101708. 2018. Xiamen.
Interview 102512x. 2018. Xiamen.
Interview 112300. 2018. Beijing.
Interview 120803. 2018. Xiamen.
Interview 171019. 2018. Beijing.
Interview 180312a. 2013. Nanchang.
Interview 181619. 2013. Beijing.
Interview 191704. 2018. Xiamen.
Interview 231908. 2018. Zhenjiang.
Interview 232407. 2013. Shenzhen.
Interview 240107a. 2013. Shenzhen.
Interview 242612. 2018. Nanchang.
Interview 250808b. 2018. Nanchang.
Personal Conversation 081724a. 2013. Düsseldorf.
Personal Conversation 241326. 2018. Beijing.
Personal Conversation 102000. 2024. Virtual.

Notes

Chapter 1

1. The Deutsche Gesellschaft für Internationale Zusammenarbeit (German Corporation for International Cooperation) has worked with national-level government bureaucratic agencies to provide a series of capacity-building programs to help Chinese subnational governments plan climate actions. These programs include *Leadership Training on Environmental and Climate Issues,* https://www.giz.de/en/worldwide/35702.html; and *Reaching China's Climate Goals Faster*, https://www.giz.de/en/worldwide/133752.html.
2. Local bureaucratic leaders who have the authority to approve lucrative projects can also be lured to engage in rent-seeking behaviors. See a case in "'Chuzhang Economy' Is a Form of Rent-Seeking ['Chuzhangjingji' shi quanli xunzu guaitai]," *Guangzhou Daily*, March 12, 2014; "Eradicate the Ground of 'Chuzhang Economy' [Chanchu 'chuzhangjingji' wotu]," *China Discipline News*, March 25, 2014.

Chapter 2

1. "Ministry of Ecology and Environment: Many Cities Still Fail to Meet Air Quality Standards, Meteorological Factors Remain Influential [Huanjingbu: reng you bushaochengshi kongqi zhiliang budabiao, zongtishang meiyou baituo qixiang yingxiang]," *Pengpai News*, September 25, 2023, https://www.thepaper.cn/newsDetail_forward_24734410.
2. "Director of the Department of Climate Change of the Ministry of Ecology and Environment: In the Next Ten Years, China Will Do its Utmost to Control Emissions [Shengtai huanjing bu qihousi sizhang: weilaishinian, women jiang jin zuidanuli kongpai]," *Beijing News* (Xinjingbao), September 25, 2019.
3. Damian Carrington, "China Makes its First Commitment to Climate Change Targets," *The Guardian*, September 22, 2009, https://www.theguardian.com/environment/2009/sep/22/hu-jintao-new-york1.
4. "Li Ganjie, China Has Fully Meet its Climate Commitments [Li Ganjie: yingdui qihoubianhua zhongguo shizhong shi baifenzhibai lüxing ziji de yiwu]," *Xinhuanet*, March 11, 2019.
5. "Zhang Xuejin, Ministry of Water Resources: Treating the Development of Small Hydropower with an Objective Attitude [Shuilibu Zhangxuejin: yongkeguantaidukandai xiaoshuidian fazhan]," *China Power News* (Zhongguo dianli bao), July 17, 2015.
6. "Department of Climate Change Completes its Institutional Reform: What's New in China's Response to Climate Change [Qihousi wancheng zhuanli: zhongguo yingdui qihoubianhua you naxie xinqixiang]," *China Youth Daily* (Zhonguo qingnian bao), November 11, 2018.

7. Steven Lee Meyers, "China's Pledge to Be Carbon Neutral by 2060: What it Means," *New York Times*, September 23, 2020, https://www.nytimes.com/2020/09/23/world/asia/china-climate-change.html.

Chapter 4

1. "Employment Expansion Should Support Small and Medium-Sized Enterprises [Kuoda jiuye xü shandai Zhongxiao qiye]," *Renmin Daily*, 2008.
2. "Seven Provinces and Cities to Launch Carbon Emissions Trading Pilots [Guonei qi shengshi jiang qi dong tan jiaoyi shidian]," *China Energy News*, November 21, 2011.
3. "Shenzhen Carbon Emissions Trading System Has Reduced More than 58 Million Tonnes of CO_2 [Tan shi chang jiao yi liang chao 5800 wan dun, shen zhen jiang ru he zhu li quan guo shi xian tan zhong he]," *Zhengquanshibao*, March 18, 2021.
4. "National Carbon Emissions Trading Capacity Building Center Settled in Shenzhen [Quan guo tan shi chang neng li jian she zhong xin luo hu shen zhen]," *China Energy News* (Zhongguo nengyuan bao), March 28, 2016, https://paper.people.com.cn/zgnyb/html/2016-03/28/content_1665161.htm.
5. The seminar was organized by the Shenzhen Municipal People's Government; Consulate General of the Kingdom of the Netherlands in Guangzhou; the Ministry of Economic Affairs, Agriculture & Innovation (the Netherlands); and the Ministry of Infrastructure and the Environment (the Netherlands). http://www.mfa.gov.cn/eng/wjb/zzjg/tyfls/tfsxw/t283051.html
6. This is a high-level cooperative framework between China and the European Union for addressing climate change, which was signed on October 19, 2006. See http://www.mfa.gov.cn/eng/wjb/zzjg/tyfls/tfsxw/t283051.html.
7. "Shenzhen Hosts the 4th International Low-Carbon City Forum [Shenzhen jüban disijie guoji ditancheng luntan]," *People's Daily*, June 18, 2016, http://paper.people.com.cn/rmrbhwb/html/2016-06/18/content_1688400.htm.

Chapter 5

1. "Nanchang's Green and Low-carbon Development Path (Lüse Nanchang de ditan fazhan zhilu)," *Beijing Weekly* (Beijing zhoubao), September 11, 2012.
2. Calculated using the exchange rate between the RMB (Renminbi) and US dollar as of 2009.
3. "Jiangxi's 6.2 Billion Local Debt Determined its Scope of Use [Jiangxi 62 yi difang zhai queding shiyong fanwei]," *Xinhuanet*, August 22, 2009.
4. "Jiangxi Establishes a Green Evaluation System [Jiangxi jianli lüse kaohe tixi]," *Guangming Daily*, January 27, 2014.
5. "Promoting Large-scale Deelopment of Green Buildings in Nanchang, Jiangxi Province [Jiangxi nanchangshi tuidong lüse jianzhu guimo hua fazhan ceji]," *China Construction News*, March 2, 2016.
6. "Confused Account of Chief Economist: Analysis of Liu Hua, Former Chief Economist of Jiangxi Nanchang Development and Reform Commission [Zongjingjishi de hutuzhang—Jiangxi nanchangfagaiwei yuanzongjingjishi liuhua yanzhong weiji wenti poxi]," *Discipline Inspection News* [Zhongguo jijian jiancha bao], December 7, 2017.

7. “One Disciplinary Case Prompts 19 to Voluntarily Return Illicit Funds [Yici anjian jingshi, 19 ren zhudong shangjiao weijikuan],” *Discipline Inspection News*, January 16, 2019.

Chapter 6

1. The National Center for Climate Change Strategy and International Cooperation, formerly affiliated with the National Development and Reform Commission, became affiliated with the Ministry of Ecology and Environment in 2018, as the Department of Climate Change moved from the National Development and Reform Commission to the Ministry of Ecology and Environment.
2. Calculated using the 2013 exchange rate between the RMB (Renminbi) and US dollar.
3. “Zhenjiang Tells the World a Low-Carbon Story [Zhenjiang: sifu lüse zhiyue xiangshijie jianghao ditan gushi],” *Yangtz Evening Post*, October 24, 2019.
4. “Building a Clean, Low-Carbon, Safe, and Efficient Energy Supply System in Zhenjiang by 2025 [Zhenjiang dao 2025 jiben xingcheng qingjie ditan anquan gaoxiao nengyuan gongying tixi],” *Zhenjiang Daily*, June 14, 2022, https://jsnews.jschina.com.cn/zj/a/202206/t20220614_3017892.shtml.

Chapter 7

1. “Xiamen Low-Carbon Economy Report: Building a Low-Carbon City, Promoting Low-Carbon Living [Xiamen ditan jingji baogao: jianshe ditan chengshi zhuzhong ditan shenghuo],” *China Environmental News* (Zhongguo huanjingbao), August 22, 2012.
2. “Xiamen Releases the ‘Beautiful Xiamen Strategic Plan’ [Meili Xiamen zhanlue quanwen fabu],” *Xiamen Daily*, March 5, 2014.
3. “Wang Menghui: Build Large-Capacity Express Transit to Advance Xiamen’s Island-to-Bay Transition [Wang Menghui: Dazao dayunliang kuaisu jiaotong tuijin haidao dao haiwan zhuanxing],” *Xiamen TV News*, December 31, 2011.
4. Calculated using the 2013 exchange rate between the RMB (Renminbi) and US dollar.
5. “Xiamen Launches Low-Carbon Pilot Demonstration [Xiamen kai zhan di tan shi dian shi fan chuang jian],” *China Environmental News*, October 25, 2023.

Chapter 8

1. “Xi Jinping Presides over the 36th Collective Study of the Politburo of the CPC Central Committee and Delivers an Important Speech [Xi Jinping zhuchi Zhonggong Zhongyang Zhengzhiju di sanshiliuci jitixuexi bing fabiao zhongyao jianghua],” Xinhua News Agency, January 25, 2022, https://www.gov.cn/xinwen/2022-01/25/content_5670359.htm.
2. Peiyuan Guo, “Opinion: Shenzhen Passes China’s First Local Green Finance Rules—Now What?” *Caixin News*, December 14, 2021.
3. Jacob Dreyer, “Xi Thinks China Can Slow Climate Change: What if He’s Right?” *New York Times*, April 19, 2024, https://www.nytimes.com/2024/04/19/opinion/china-climate-xi-jinping.html.
4. “Pursuing Carbon Peaking and Carbon Neutrality Actively and Prudently [Ji ji wen tuo tui jin tan da feng he tan zhong he],” *Renmin Daily*, April 6, 2023.

5. This group was created in 2021, after Xi's carbon peak and carbon neutrality pledges in 2020, to coordinate the implementation of climate goals. It was quietly dissolved in 2023, partly due to China's government restructuring that year given the existence of the National Leading Group on Climate Change, Energy Conservation, and Emissions Reduction, which was established in 2007.
6. "Xie Zhenhua: The 1 + N Policy System for Carbon Peaking and Carbon Neutrality to be Released Soon, Aiming to Accelerate Transformation Across Ten Key Areas [Xie Zhenhua: tan da feng zhong he 1+N zheng ce ti xi ji jiang fa bu shi ge ling yü jia su zhuan xing]," *Beijing News*, July 24, 2021.
7. We Are Still In, "'We Are Still in' Declaration," 2018, https://www.wearestillin.com/we-are-still-declaration.
8. Office of Governor Gavin Newsom, "What Governor Newsom's Trip to China Accomplished," October 30, 2023, https://www.gov.ca.gov/2023/10/30/what-governor-newsoms-trip-to-china-accomplished/. The Memoranda of Understanding documents can be found on the California Energy Commission's website: https://www.energy.ca.gov/about/international-cooperation/climate-change-partnerships.

Bibliography

Anderton, Karen, and Joana Setzer. 2018. "Subnational Climate Entrepreneurship: Innovative Climate Action in California and São Paulo." *Regional Environmental Change* 18: 1273–1284. https://doi.org/10.1007/s10113-017-1160-2.

Ang, Yuen Yuen. 2016. *How China Escaped the Poverty Trap*. Cornell University Press.

Baeumler, Axel, Ede Ijjasz-Vasquez, and Shomik Mehndiratta. 2012. "Sustainable Low-Carbon Cites in China: Why it Matters and What Can be Done." In *Sustainable Low-Carbon City Development in China*, edited by Axel Baeumler, Ede Ijjasz-Vasquez and Shomik Mehndiratta. Springer.

Baumgartner, Frank. R., and Bryan D. Jones. 2009. *Agenda and Instability in American Politics*. 2nd ed. University of Chicago Press.

Betsill, Michele M., and Harriet Bulkeley. 2006. "Cities and the Multilevel Governance of Global Climate Change." *Global Governance* 12, no. 2: 141–159.

"Building a Clean, Low-Carbon, Safe, and Efficient Energy Supply System in Zhenjiang by 2025. [Zhenjiang dao 2025 jiben xingcheng qingjie ditan anquan gaoxiao nengyuan gongying tixi]." 2022. *Zhenjiang Daily*, June 14. https://jsnews.jschina.com.cn/zj/a/202206/t20220614_3017892.shtml.

Bulkeley, Harriet. 2013. *Cities and Climate Change*. Routledge.

Bulkeley, Harriet. 2023. "The Condition of Urban Climate Experimentation." *Sustainability: Science, Practice and Policy* 19, no. 1: 2188726. https://doi.org/10.1080/15487733.2023.2188726.

Bulkeley, Harriet, Vanesa Castan Broto, and Gareth A. S. Edwards. 2015. *An Urban Politics of Climate Change: Experimenting and the Governing of Socio-Technical Transitions*. Routledge.

Bulman, David J. 2016. *Incentivized Development in China: Leaders, Governance, and Growth in China's Counties*. Cambridge University Press.

California Energy Commission. 2025. *Climate Change Partnership*. California Energy Commission. https://www.energy.ca.gov/about/international-cooperation/climate-change-partnerships.

Campbell, Kurt M., and Jake Sullivan. 2019. "Competition Without Catastrophe: How America Can Both Challenge and Coexist with China." *Foreign Affairs*, August 1. https://www.foreignaffairs.com/china/competition-with-china-catastrophe-sullivan-campbell.

Cao, Xun, Genia Koskta, and Xu Xu. 2019. "Environmental Political Business Cycles: The Case of PM2.5 Air Pollution in Chinese Prefectures." *Environmental Science & Policy* 93: 92–100.

Carrington, Damian. 2009. "China Makes its First Commitment to Climate Change Targets." *The Guardian*, September 22. https://www.theguardian.com/environment/2009/sep/22/hu-jintao-new-york1.

Cashore, Benjamin, Steven Bernstein, David Humphreys, Ingrid Visseren-Hamakers, and Katharine Rietig. 2019. "Designing Stakeholder Learning Dialogues for Effective Global Governance." *Policy and Society* 38, no. 1: 118–147. https://doi.org/10.1080/14494035.2019.1579505.

Chen, Ling. 2018. *Manipulating Globalization: The Influence of Bureaucrats on Business in China*. Stanford University Press.

Chen, Xing, and Jintao Xu. 2018. "Carbon Trading Scheme in the People's Republic of China: Evaluating Performance of Seven Pilot Projects." *Asian Development Review* 35, no. 2: 131–152.

Chen, Ying Chun, Delin Wu, Jun Liu, Jiahua Qiao, Yong Qian, and Xinpin Qian. 2013. *Report on Shenzhen's Financial and Economic Development, 2001–2010* [Shenzhen shi cai shui jing ji shi nian fa zhan bao gao 2001–2010]. China Financial & Economic Publishing House.

Cheshmehzangi, Ali, Linjun Xie, and May Tan-Mullins. 2018. "The Role of International Actors in Low-Carbon Transitions of Shenzhen's International Low Carbon City in China." *Cites* 74: 64–74.

Chien, Shiuh-Shen, and Dong-Li Hong. 2018. "River Leaders in China: Party-State Hierarchy and Transboundary Governance." *Political Geography* 62: 58–67.

China Council for International Cooperation on Environment and Development. 2019. *Special Policy Study Report: Global Climate Governance and China's Role.* https://cciced.eco/wp-content/uploads/2020/06/cciced-sps-1-1-climate.pdf.

"'Chuzhang Economy' Is a Form of Rent-Seeking [Chuzhangjingji' shi quanli xunzu guaitai]." 2014. *Guangzhou Daily*, March 12.

Chung, Jae Ho. 2000. *Central Control and Local Discretion in China: Leadership and Implementation During Post-Mao Decollectivization.* Oxford University Press.

"Confused Account of Chief Economist: Analysis of Liu Hua, Former Chief Economist of Jiangxi Nanchang Development and Reform Commission [Zongjingjishi de hutuzhang—Jiangxi nanchangfagaiwei yuanzongjingjishi liuhua yanzhong weiji wenti poxi]." 2017. *Discipline Inspection News* (Zhongguo jijian jiancha bao), December 7.

Culhane, Trevor, Galen Hall, and J. Timmons Roberts. 2021. "Who Delays Climate Action? Interest Groups and Coalitions in State Legislative Struggles in the United States." *Energy Research & Social Science* 79: 1021114.

Dauvergne, Peter. 2018. *Environmentalism of the Rich.* MIT Press.

Davidson, Michael R., Valerie J. Karplus, Joanna I. Lewis, Jonas Nahm, and Alex Wang. 2022. "Risks of Decoupling from China on Low-Carbon Technologies." *Science* 377, no. 6612: 1266–1269. https://doi.org/10.1126/science.abq5446.

Delman, Jørgen. 2016. "Urban Climate Change Politics in China: Fragmented Authoritarian and Governance Innovations in Hangzhou." In *Chinese Politics as Fragmented Authoritarianism*, edited by Kjeld Erik Brødsgaard. Routledge.

"Department of Climate Change Completes its Institutional Reform: What's New in China's Response to Climate Change [Qihousi wancheng zhuanli: Zhongguo yingdui qihou bianhua you naxie xinqixiang]." 2018. *China Youth Daily* (Zhongguo Qingnian Bao), November 11.

Department of Ecology and Environment of Jiangsu Province. 2020. *The Provincial Environmental Protection Inspection Team Reported Findings to Changzhou, Huai'an, and Zhenjiang.* November 2. https://sthjt.jiangsu.gov.cn/art/2020/11/2/art_83552_10044472.html.

Development and Reform Commission of Shenzhen Municipality. 2011a. *Regarding the Submission of the Report on the Request for Support for the Construction of the China–The Netherlands (Europe) Low-Carbon City Project in Shenzhen* [Guanyü shangbao ken qing zhichi shenzhen jianshe zhong he ou ditancheng xiang mu dai ni gao de qing shi].

Development and Reform Commission of Shenzhen Municipality. 2011b. *Report on the China–The Netherlands (Europe) Low-Carbon City Seminar* [Guan yü zhong he ou di tan cheng zhuan jia yan tao hui qing kuang de bao gao].

Development and Reform Commission of Shenzhen Municipality. 2011c. *Report on the Meeting with Director Su Wei from the National Development and Reform Commission's*

Department of Climate Change [Guanyü bai jian guojia fazhan gaigeweiying dui qihou bianhua si su wei si zhang qing kuang de baogao].

Development and Reform Commission of Shenzhen Municipality. 2012. *Shenzhen Low-Carbon Pilot City Work Implementation Plan* [Shenzhen shi ditan cheng shi shi dian gongzuo shishifang an]. December 29.

Development and Reform Commission of Shenzhen Municipality. 2020. *Shenzhen New Renewable Vehicle Development Work Plan (2021–2025)* [Shenzhen shixin nengyuanqiche tui guangyingyong gongzuofangan (2021–2025)]. https://fgw.sz.gov.cn/zwgk/zcjzcjd/zc/content/post_8667568.html.

Ding, Iza. 2020. "Performative Governance." *World Politics* 72, no. 4: 525–556.

Ding, Iza. 2022. *The Performative State: Public Scrutiny and Environmental Governance in China*. Cornell University Press.

Ding, Iza, and van der Kamp, Denise. 2020. "High Maintenance or Low Maintenance? Environmental Policy Implementation in China." Paper presented at the 2020 Annual Meeting of American Political Science Association, September 11–13.

"Director of the Department of Climate Change of the Ministry of Ecology and Environment: In the Next Ten Years, China Will Do its Utmost to Control Emissions [Sheng tai huan jing bu qi hou si si zhang: wei lai shi nian, wo guo jiang jin zui da nu li kong pai]." 2019. *Beijing News* (Xinjingbao), September 24.

Doig, Jameso W., and Erwin C. Hargrove. 1990. "Leadership and Political Analysis." In *Leadership and Innovation: Entrepreneurs in Government*, edited by Jameson W. Doig and Erwin C. Hargrove. Johns Hopkins University Press.

Dong, Wenjuan, and Ye Qi. 2015. *Government Behavior and PPPs (Public–Private Partnerships) in the Development of China's Clean Energy Industry* [Zhong guo qing jie neng yuan chan ye fa zhan zhong de zheng fu xing wei yü PPP mo shi]. Policy Report Series. Brookings-Tsinghua Ceter for Public Policy.

Dreyer, Jacob. 2024. "Xi Thinks China Can Slow Climate Change. What If He's Right?" *The New York Times*, April 19. https://www.nytimes.com/2024/04/19/opinion/china-climate-xi-jinping.html.

Dubash, Navroz K., Aditya Valiathan Pillai, Christian Flachsland, et al. 2021. "National Climate Institutions Complement Targets and Policies." *Science* 374, 690—693. DOI:10.1126/science.abm1157.

Eaton, Sarah, and Genia Kostka. 2013. "Does Cadre Turnover Help or Hinder China's Green Rise? Evidence from Shanxi Province." In *Chinese Environmental Governance: Dynamics, Challenges, and Prospects in a Changing Society*, edited by Bingqian Ren, and Huisheng Shou. Palgrave Macmillan.

Eaton, Sarah, and Genia Kostka. 2014. "Authoritarian Environmentalism Undermined? Local Leaders' Time Horizons and Environmental Policy Implementation in China." *China Quarterly* 218: 359–380.

Economy, Elizabeth C. 2004. *The River Runs Black: The Environmental Challenge to China's Future*. Cornell University Press.

Economy, Elizabeth C. 1997. "Chinese Policy-Making and Global Climate Change: Two-Front Diplomacy and the International Community." In *The Internationalization of Environmental Protection*, edited by Miranda A. Schreurs and Elizabeth C. Economy. Cambridge University Press.

Edwards, Bob, and John D. McCarthy. 2007. "Resources and Social Movement Mobilization." In *The Blackwell Companion to Social Movement*, edited by David A. Snow, Sarah A. Soule, and Hanspeter Kriesi. Blackwell Publishing.

"Employment Expansion Should Support Small and Medium-Sized Enterprises [Kuodajiuye xü shandai zhongxiaoqiye]." 2008. *Renmin Daily*, December 25.

"Eradicate the Ground of 'Chuzhang Economy' [Chanchu 'chuzhangjingji' wotu]." 2014. *China Discipline News*, March 25.

Fewsmith, Joseph. 2013. *The Logic and Limits of Political Reform in China.* Cambridge University Press.

Filimonova, Nadezhda. 2024. "Constructing Climate Change: Exploring How Cities Frame Climate Change in the Arctic." *Journal of Urban Affairs*, 1–19. https://doi.org/10.1080/07352166.2024.2351391.

Florini, Ann M., Hairong Lai, and Yeling Tan. 2012. *China Experiments: From Local Innovations to National Reform.* Brookings Institution Press.

Frisch-Aviram, Neomi, Nissim Cohen, and Itai Beeri. 2018. "Low-Level Bureaucrats, Local Government Regimes and Policy Entrepreneurship." *Policy Sciences* 51: 39–57. https://doi.org/10.1007/s11077-017-9296-y.

Fujian Industry and Information Technology Department (Fujian sheng gongye he xinxi hua ting). 2019. *Notice of the Fujian Provincial Department of Industry and Information Technology on the Release of the List of Benchmarking Enterprises and Key Projects of the Industrial Internet Demonstration Platform* [Fujian sheng gongye he xinxi hua ting guanyü fa bu hulianwang shi fan ping tai biao gan qiye ji zhong dian xiang mu ming dan de tongzhi]. December 25.

Gallagher, Kelly Sims. 2014. *The Globalization of Clean Energy Technology: Lessons from China.* MIT Press.

Ganz, Marshall. 2000. "Resources and Resourcefulness: Strategic Capacity in the Unionization of California Agriculture, 1959–1966." *American Journal of Sociology* 105, no. 4: 1003–1062.

General Office of Nanchang Municipal Government (Nanchang shi ren min zheng fu ban gong ting). 2012. *Notice of Nanchang Municipal Government's Opinions on Further Supporting the Ecological Protection and Scientific Development of Wanli District* [Nanchang shi renminzheng fu yin fa guanyü jin yi bu zhi chi wan li qü shengtaibaohu, kexuefazhan de yijian de tongzhi]. March 21.

General Office of Shenzhen Municipal Committee of the Communist Party of China (Shenzhen shi shiweibangong ting). 2012. *Overarching Approach to Building the International Low-Carbon City* [Guo ji di tan cheng gui hua jian she zong ti si lu].

General Office of Shenzhen Municipal People's Government (Shenzhen shi ren min zhengfu ban gong ting). 2011. *Minutes of the Seminar on China–The Netherlands (Europe) Low-Carbon City Cooperation and Related Issues* [Guan yü yan jiu zhong he (Ouzou) di tan cheng you guan wen ti de hui yi ji yao].

General Office of Shenzhen Municipal People's Government (Shenzhen shi ren min zhengfu ban gong ting). 2012a. *Minutes of the City Leadership Office Conference No. 28* [Shi ling dao ban gong hui yi ji yao 2012 28 hao].

General Office of Shenzhen Municipal People's Government (Shenzhen shi ren min zhengfu ban gong ting). 2012b. *Minutes of Shenzhen International Low-Carbon City Planning and Development Conference 387* [Shenzhen guo ji di tan cheng gui hua jian she hui yi ji yao 387].

General Office of Shenzhen Municipal People's Government (Shenzhen shi ren min zhengfu ban gong ting). 2012c. *Minutes of Shenzhen International Low-Carbon City Planning and Development Conference 246* [Shenzhen guo ji di tan cheng gui hua jian she hui yi ji yao 246].

General Office of Shenzhen Municipal People's Government (Shenzhen shi ren min zhengfu ban gong ting). 2012d. *Minutes of Shenzhen International Low-Carbon City Planning and Development Conference 338* [Shenzhen guo ji di tan cheng gui hua jian she hui yi ji yao 338].

General Office of Xiamen Municipal Government (Xiamen shi ren min zhengfu bangong ting). 2017. *Notice of the General Office of the Xiamen Municipal People's Government on Issuing and Implementing the Implementation Plan for the Construction of the Carbon Emissions Trading Market* [Xiamen shi ren min zheng fu ban gong ting guan yü yin fa guan che luo shi tan pai fang quan jiao yi shi chang jian she shi shi fang an de tong zhi]. May 18.

General Office of Zhenjiang Municipal Government. 2021. *2021 Zhenjiang Ecological Civilization (Low-Carbon City) Construction Plan* [Guan yü yin fa 2021 quan shi sheng tai wen ming (Di tan cheng shi) jian she jiu da xing dong ren wu de tong zhi].

Gilley, Bruce. 2012. "Authoritarian Environmentalism and China's Response to Climate Change." *Environmental Politics* 21, no. 2: 287–307.

Gilley, Bruce. 2017. "Local Governance Pathways to Decarbonization in China and India." *China Quarterly* 231: 728–748.

GIZ (German Corporation for International Cooperation). *Leadership Training on Environmental and Climate Issues.* Accessed June 29, 2025. https://www.giz.de/en/worldwide/35702.html.

GIZ. *Reaching China's Climate Goals Faster.* Accessed June 29, 2025. https://www.giz.de/en/worldwide/133752.html.

Gong, Weila. 2022. "Temporary Leaders and Stable Institutions: How Local Bureaucratic Entrepreneurs Institutionalize China's Low-Carbon Policy Experiments." *China Quarterly* 252: 1206–1232. https://doi.org/10.1017/S0305741022000790.

Gong, Weila, and Joanna I. Lewis. 2023. "The Role of International Engagement in Greening China's Belt and Road Initiative." *Environmental Politics* 32, no. 7: 1208–1230. https://doi.org/10.1080/09644016.2023.2223074.

Gordon, David J. 2020. *Cities on the World Stage: The Politics of Global Urban Climate Governance.* Cambridge University Press.

Goyal, Nihit, Michael Howlett, and Namrata Chindarkar. 2020. "Who Coupled Which Stream(s)? Policy Entrepreneurship and Innovation in the Energy–Water Nexus in Gujarat, India." *Public Administration and Development* 40, no. 1: 49–64. https://doi.org/10.1002/pad.1855.

Green, Jessica F. 2017. "Policy Entrepreneurship in Climate Governance: Toward a Comparative Approach." *Environment and Planning C: Politics and Space* 35, no. 8: 1471–1482. https://doi.org/10.1177/2399654417735905.

Green, Jessica F., and Raúl Salas Reyes. 2023. "The History of Net Zero: Can We Move from Concepts to Practice?" *Climate Policy* 23, no. 7: 901–915. https://doi.org/10.1080/14693062.2023.2218334.

Grossman, Gene M., and Alan B. Krueger. 1995. "Economic Growth and the Environment." *Quarterly Journal of Economics* 110, no. 2: 353–377. https://doi.org/10.2307/2118443.

Guo, Peiyuan. 2021. "Opinion: Shenzhen Passes China's First Local Green Finance Rules—Now What?." *Caixin News*, December 14.

Guttman, Dan, Oran Young, Yijia Jing, Barbara Bramble, Maoling Bu, Carmen Chen et al. 2018. "Environmental Governance in China: Interactions between the State and 'Nonstate Actors.'" *Journal of Environmental Management* 220: 126–135.

Habich-Sobiegalla, Sabrina. 2018. "How Do Central Control Mechanisms Impact Local Water Governance in China? The Case of Yunnan Province." *China Quarterly* 234: 444–462.

Hasmath, Reza, Jessica C. Teets, and Orion A. Lewis. 2019. "The Innovative Personality? Policy Making and Experimentation in an Authoritarian Bureaucracy." *Public Administration and Development* 39, no. 3: 154–162.

He, Alex Jingwei. 2018. "Maneuvering within a Fragmented Bureaucracy: Policy Entrepreneurship in China's Local Healthcare Reform." *China Quarterly* 236: 1088–1110.

Heffer, Abbey S., and Gunter Schubert. 2023. "Policy Experimentation under Pressure in Contemporary China." *China Quarterly* 253: 35–56. https://doi.org/10.1017/S0305741022001801.

Heggelund, Gørild. 2007. "China's Climate Change Policy: Domestic and International Developments." *Asian Perspective* 31, no. 2: 155–191.

Heilmann, Sebastian. 2008a. "From Local Experiments to National Policy: The Origins of China's Distinctive Policy Process." *China Journal* 59: 1–30.

Heilmann, Sebastian. 2008b. "Policy Experimentation in China's Economic Rise." *Studies in Comparative International Development* 43: 1–26.

Heilmann, Sebastian, and Elizabeth J. Perry. 2011. "Embracing Uncertainty: Guerrilla Policy Style and Adaptive Governance in China." In *Mao's Invisible Hand: The Political Foundations of Adaptive Governance in China*, edited by Sebastian Heilmann, and Elizabeth J. Perry. Harvard University Asia Center.

Helco, Hugo. 1978. "Issue Networks and the Executive Establishment." In *The New American Political System*, edited by Anthony King. American Enterprise Institute.

Hoffmann, Matthew J. 2011. *Climate Governance at the Crossroads: Experimenting with a Global Response after Kyoto*. Oxford University Press.

Hopper, JoyAnna. S. and Clint. S. Swift. 2021. "Considering Subnational Support of Climate Change Policy in the United States and the Implications of Symbolic Policy Acts." *Environmental Politics* 31, no. 6: 969–990. https://doi.org/10.1080/09644016.2021.1964771.

Hsu, Angel, Amy J. Weinfurter, and Kaiyang Xu. 2017. "Aligning Subnational Climate Actions for the New Post-Paris Climate Regime." *Climatic Change* 142, no. 3: 419–432. https://doi.org/10.1007/s10584-017-1957-5.

Huang, Xiaochun, and Li-an Zhou. 2023. "'Paired Competition': A New Mechanism for the Innovation of Urban Grassroots Governance." *Chinese Journal of Sociology* 9, no. 1: 3–45.

Hughes, Sara. 2019. *Repowering Cities: Governing Climate Change Mitigation in New York City, Los Angeles, and Toronto*. Cornell University Press.

Huitema, Dave, Elin Lerum Boasson, and Raoul Beunen. 2018. "Entrepreneurship in Climate Governance at the Local and Regional Levels: Concepts, Methods, Patterns, and Effects." *Regional Environmental Change* 18: 1247–1257. https://doi.org/10.1007/s10113-018-1351-5.

Hultman, Nathan E., Leon Clarke, Carla Frisch, et al. 2020. "Fusing Subnational with National Climate Action Is Central to Decarbonization: The Case of the United States." *Nature Communication* 11: 5255. https://doi-org.libproxy.berkeley.edu/10.1038/s41467-020-18903-w.

Ikenberry, G. John. 1988. *Reasons of State: Oil Politics and the Capacities of American Government*. Cornell University Press.

Institute of Urban Environment, Chinese Academy of Sciences. *Evaluating Key Area Functions and Environmental Impact in Southern Haicang District*, Xiamen. 2007. Xiamen: Institute of Urban Environment, Chinese Academy of Sciences, Xiamen.

International Energy Agency. 2021. *An Energy Sector Roadmap to Carbon Neutrality in China*. https://www.iea.org/reports/an-energy-sector-roadmap-to-carbon-neutrality-inchina/executive-summary.

International Energy Agency. 2023. *Renewables 2023*. https://www.iea.org/reports/renewables-2023/executive-summary.

International Energy Agency. 2024. "China: Coal." https://www.iea.org/countries/china/coal.

Jahiel, Abigail R. 1998. "The Organization of Environmental Protection in China." *China Quarterly* 156: 757–787.

Jänicke, Martin. 2012. *Megatrend Environmental Innovation: Towards Ecological Modernization from Business and the State* [Megatrend Umweltinnoation: Zur oekologischen Modernisierng von Wirtschaft und Staat]. 2nd ed. Oekom.

Jänicke, Martin, and Klaus Jacob. 2004. "Lead Markets for Environmental Innovations: A New Role for the Nation State." *Global Environmental Politics* 4, no. 1: 29–46.

Jaros, Kyle A., and Sara A. Newland. 2024. "Paradiplomacy in Hard Times: Cooperation and Confrontation in Subnational U.S.–China Relations." *Publius: The Journal of Federalism* 54, no. 4: 599–627.

"Jiangxi Establishes a Green Evaluation System [Jiangxi jianli luse kaohe tixi]." 2014. *Guangming Daily*, January 27.

"Jiangxi's 6.2 Billion Local Debt Determined its Scope of Use [Jiangxi: 62 yiyuan 2009 nian difang zhengfu zhaiquan que ding shiyong fanwei]." 2009. *Xinhuanet*, August 22.

Jiang, Shuxia. 2009. *Local Government during the Reform and Opening-up Era: Xiamen's Three-Decade Transformation* [Gai ge kai fang zhong de di fang zheng fu: xia men bian qian 30 nian biao ben kao cha]. Truth & Wisdom Press.

Jiang, Wei. 2015. "Could a Less Developed City Solve its CO2 Emission Dilemma? Evidence from a Low Carbon Pilot City." *Chinese Journal of Urban and Environmental Studies* 3, no. 1: 1550007.

Jiangxi Housing and Urban–Rural Development Commission. 2020. *Announcement on the 2019 Provincial Progress in Building Energy Efficiency and Green Building Development* [Guanyü quan sheng 2019 nian jieneng yü lüse jianzhu jin zhan de tong bao].

Jiangxi Provincial People's Congress. 2017. *The Nanchang Municipal Regulations on the Promotion of Low-Carbon Development* [Nanchang shi ditan fazhan cujin tiaoli]. June 8. http://jxrd.jxnews.com.cn/system/2017/02/16/015773016.shtml.

Jiangxi Provincial Audit Office of the People's Republic of China. 2017. *Gan Audit Report*.

Jiangxi Provincial Energy Bureau. 2020. *Bulletin on the 2019 Regional Energy Intensity Reductions and Indicators* [Guanyü 2019 nian quansheng he geshequ shi wanquan diqü shengchan zongzhi nenghao jiangdi lü deng zhibiao gongbao].

Jörgensen, Kirsten, Anu Jogesh, and Arabinda Mishra. 2015. "Multi-Level Climate Governance and the Role of the Subnational Level." *Journal of Integrative Environmental Sciences* 12, no. 4: 235–245. https://doi.org/10.1080/1943815X.2015.1096797.

Jones, Bryan D. 1989. "Causation, Constraint, and Political Leadership." In *Leadership and Politics: New Perspectives in Political Science*, edited by Bryan D. Jones. University Press of Kansas.

Karplus, Valerie J., M. Granger Morgan, and David G. Victor. 2021. "Finding Safe Zones for Science." *Issues in Science and Technology* 38, no. 1: 76–81. https://www.jstor.org/stable/27092120.

Kelsey, Nina. 2021. "Cities, Commons, and the Unilateral Provision of Public Goods." *International Relations* 35, no. 3: 489–509. https://doi.org/10.1177/00471178211037089.

Kennedy, John James, and Dan Chen. 2014. "Election Reform from the Middle and at the Margins." In *Local Governance Innovation in China: Experimentation, Diffusion and Defiance*, edited by Jessica C. Teets and William Hurst. Routledge.

Kern, Kristine. 2018. "Cities as Leaders in EU Multilevel Climate Governance: Embedded Upscaling of Local Experiments in Europe." *Environmental Politics* 28, no. 1: 125–145. https://doi.org/10.1080/09644016.2019.1521979.

Khanna, Nina, David Fridley, and Lixuan Hong. 2014. "China's Pilot Low-Carbon City Initiative: A Comparative Assessment of National Goals and Local Plans." *Sustainable Cities and Society* 12: 110–121.

Kingdon, John W. 1995. *Agendas, Alternatives and Public Policies*. 2nd ed. Longman.

Koehn, Peter H. 2015. *China Confronts Climate Change: A Bottom-Up Perspective*. Routledge.

Kostka, Genia, Xuehua Zhang, and Kyoung Shin. 2019. "Information, Technology, and Digitalization in China's Environmental Governance." *Journal of Environmental Planning and Management* 63, no. 1: 1–13.

Kou, Chien-wen, and Wen-Hsuan Tsai. 2014. "'Sprinting with Small Steps' Towards Promotion: Solutions for the Age Dilemma in the CCP Cadre Appointment System." *China Journal* 71: 153–171. https://doi.org/10.1086/674558.

Lampton, David M. 1992. "A Plum for a Peach: Bargaining, Interest, and Bureaucratic Politics in China." In *Bureaucracy, Politics, and Decision Making in Post-Mao China*, edited by David M. Lampton and Kenneth G. Lieberthal. University of California Press.

Lee, Henry, Daniel P. Schrag, Matthew Bunn, Michael Davidson, Wei Peng, Wang Pu et al. 2021. *Foundations for a Low-Carbon Energy System in China*. Cambridge University Press.

Lewis, Joanna I. 2012. *Green Innovation in China: China's Wind Power Industry and the Global Transition to a Low-Carbon Economy*. Columbia University Press.

Lewis, Joanna I., David G. Fridley, Lynn K. Price, Hongyou Lu, and John P. Romankiewicz. 2015. "Understanding China's Non–Fossil Energy Targets: Methodology Standardization Will Improve Comparability." *Science* 350, no. 6264: 1034–1036.

Lewis, O., J. Teets, and R. Hasmath 2022. "Exploring Political Personalities: The Micro-Foundation of Local Policy Innovation in China." *Governance* 35, no. 1: 103–122.

"Li Ganjie: China Has Fully Met its Climate Commitments [Li Ganjie: ying dui qi hou bian hua zhong guo shi zhong shi bai fen zhi bai di lu xing zi ji de yi wu]." 2019. *Xinhuanet*, March 11.

Li, Junfeng. 2019. "The Development and Achievements of China's Renewable Energy in the Past 70 Years [Wo guo ke zai sheng neng yuan 70 nian fa zhan li cheng yü cheng jiu]." *Energy Information Research*, September 26.

Li, Junfeng, Fengbo Cai, Liming Qiao, Hongwen Xie, Hu Gao, and Xiaosheng Yang et al. 2012. "*China's Wind Power Development Outlook 2012.*" *China Environmental Science Press*.

Li, Keqiang. 2014. "Report on the Work of the Government." Xinhua News Agency, March 14. https://www.gov.cn/guowuyuan/2014-03/14/content_2638989.htm.

Li, Lianjiang, and Kevin J. O'Brien. 2008. "Protest Leadership in Rural China." *China Quarterly* 193: 1–23.

Li, Mengnan, Haiyi Ye, Xiaowei Liao, Junping Li, and Xiaoming Ma. 2020. "How Shenzhen, China Pioneered the Widespread Adoption of Electric Vehicles in a Major City: Implications for Global Implementation." *Wires Energy and Environment* 9, no. 4. https://doi.org/10.1002/wene.373.

Li, Rong, and Zerong Xiong. 2010. "Analysis of Nanchang's Industrial Structure Adjustment under the Background of Poyang Lake Ecological Economic Zone [Po yang hu sheng tai jing ji qü bei jing xia nan chang chan ye jie gou tiao zheng fen xi]." *Journal of the Party School of CPC Nanchang Municipal Committee* 5: 53–57.

Li, Yifei, and Judith Shapiro. 2020. *China Goes Green: Coercive Environmentalism for a Troubled Planet*. Polity.

Li, Zhiqing. 2020. "Environmental Protection and Economic Development: Historical Review and Future Prospects." *World Environment* 1, no. 182: 66–70. http://www.chinaeol.net/zyzx/sjhjzz/zzlm/rdgz/202004/P020200420535379808702.pdf.

Lieberthal, Kenneth, and Michel Oksenberg. 1988. *Policy Making in China: Leaders, Structures, and Processes*. Princeton University Press.

Lin, Tao, Xiongzhi Xue, Longyu Shi, and Lijie Gao. 2013. "Urban Spatial Expansion and its Impacts on Island Ecosystem Services and Landscape Pattern: A Case Study of the Island City of Xiamen, Southeast China." *Ocean & Coastal Management* 81: 90–96.

Liu, Kun. 2019. "State Council's Report on Expenses on the Ecological and Environmental Protection." December 25. http://www.npc.gov.cn/npc/c2/c30834/201912/t20191225_303979.html.

Liu, Hengwei, Ding Ding, and Huaqing Xu. 2014. *Report on Zhenjiang's Low-Carbon Pilot City Development* [Zhenjiang shi guojia ditancheng shi hi dian gongzuo diaoyan baogao]. Beijing: National Center for Climate Change Strategy and International Cooperation.

Liu, Lei, Pu Wang, and Tong Wu. 2017. "The Role of Nongovernmental Organizations in China's Climate Change Governance." *WIREs Climate Change* 8, no. 6: e483. https://doi.org/10.1002/wcc.483.

Liu, Zhu, Zhu Deng, Gang He, Hailin Wang, Xian Zhang, Jiang Lin et al. 2022. "Challenges and Opportunities for Carbon Neutrality in China." *Nature Reviews Earth & Environment* 3: 141–155. https://doi.org/10.1038/s43017-021-00244-x.

Lo, Kevin, He Li, and Kang Chen. 2019. "Climate Experimentation and the Limits of Top-down Control: Local Variation of Climate Pilots in China." *Journal of Environmental Planning and Management* 63, no. 1: 109–126. https://doi.org/10.1080/09640568.2019.1619539.

Lo, Kevin, Mark Y. Wang, Bo Qin, and Sun Sheng Han. 2015. "Low Carbon City Development in China." In *Towards Low Carbon Cities in China: Urban Form and Greenhouse Gas Emissions*, edited by Sun Sheng Han, Ray Green, and Mark Y. Wang. Routledge.

Ma, Liang. 2017. "Site Visits, Policy Learning, and the Diffusion of Policy Innovation: Evidence from Public Bicycle Programs in China." *Journal of Chinese Political Science* 22: 581–599. https://doi.org/10.1007/s11366-017-9498-3.

Mai, Qianqing, and Maria Francesch-Huidobro. 2015. *Climate Change Governance in Chinese Cities*. Routledge.

Manion, Melanie. 2015. *Information for Autocrats: Representation in Chinese Local Congresses*. Cambridge University Press.

Meckling, Jonas, and Jonas Nahm. 2021. "Strategic State Capacity: How States Counter Opposition to Climate Policy." *Comparative Political Studies* 55, no. 3: 493–523. https://doi.org/10.1177/00104140211024308.

Mertha, Andrew. 2008. *China's Water Warriors: Citizen Action and Policy Change*. Cornell University Press.

Mertha, Andrew. 2009. "'Fragmented Authoritarianism 2.0': Political Pluralization in the Chinese Policy Process." *China Quarterly* 200: 995–1012.

Millar, Heather, Eve Bourgeois, Steven Bernstein, and Matthew Hoffmann. 2020. "Self-Reinforcing and Self-Undermining Feedbacks in Subnational Climate Policy Implementation." *Environmental Politics* 30, no. 5: 791–810. doi:10.1080/09644016.2020.1825302.

Ministry of Ecology and Environment. 2019. *2019 Report on the State of China's Ecology and Environment* [Zhongguo sheng tai huan jing gong bao 2019]. June 2.

Ministry of Ecology and Environment. 2022. *China's Polices and Actions toward Addressing Climate Change 2022*.

Ministry of Ecology and Environment. 2023a. *China's Polices and Actions toward Addressing Climate Change 2023*.

Ministry of Ecology and Environment. 2023b. *Progress Report on National Low-Carbon Pilot Cities* [Guojia ditan shidian chengshi gongzuo jinzhan pinggu baogao]. https://www.mee.gov.cn/ywgz/ydqhbh/wsqtkz/202307/W020230713602785966247.pdf.

Ministry of Ecology and Environment, National Development and Reform Commission, Ministry of Industry and Information Technology, Ministry of Housing and Urban–Rural Development, People's Bank of China, State-owned Assets Supervision and Administration Commission of the State Council, National Government Offices Administration, China Banking and Insurance Regulatory Commission, and China Securities Regulatory Commission. 2022. *Notice on the Announcement of the Climate Investment and Financing Pilot Areas* [Guanyü gongbu qihou rongzi shidian mingdan de tongzhi]. https://www.mee.gov.cn/xxgk2018/xxgk/xxgk04/202208/t20220810_991388.html.

"Ministry of Ecology and Environment: Many Cities Still Fail to Meet Air Quality Standards, Meteorological Factors Remain Influential [Huanjingbu: reng you bushaochengshi kongqi zhiliang budabiao, zongtishang meiyou baituo qixiang yingxiang]." 2023. *Pengpai News*, September 25. https://www.thepaper.cn/newsDetail_forward_24734410.

Ministry of Finance of People's Republic of China. 2020. *Notice on the Draft Revised Measures for the China Clean Development Mechanism Fund (For Public Comment)* [Zhong guo qing jie fa zhanji zhi ji jin guan li ban fa xiu ding zheng qiu yi jian]. July 13. http://www.gov.cn/hudong/2020-07/13/content_5526384.htm.

Mintrom, Michael, and Phillipa Norman. 2009. "Policy Entrepreneurship and Policy Change." *Policy Studies Journal* 37, no. 4: 649–667. https://doi.org/10.1111/j.1541-0072.2009.00329.x.

Moore, Scott M. 2014. "Modernisation, Authoritarianism, and the Environment: The Politics of China's South–North Water Transfer Project." *Environmental Politics* 23, no. 6: 947–964. https://doi.org/10.1080/09644016.2014.943544.

Morris, Aldon D., and Suzanne Staggenborg. 2007. "Leadership in Social Movements." In *The Blackwell Companion to Social Movement*, edited by David A. Snow, Sarah A. Soule and Hanspeter Kriesi. Blackwell Publishing.

Myers, Steven Lee. 2020. "China's Pledge to Be Carbon Neutral by 2060: What it Means." *New York Times*, September 23. https://www.nytimes.com/2020/09/23/world/asia/china-climate-change.html.

Nahm, Jonas. 2017. "Exploiting the Implementation Gap: Policy Divergence and Industrial Upgrading in China's Wind and Solar Sectors." *China Quarterly* 231: 705–727.

Nahm, Jonas. 2021. *Collaborative Advantage: Forging Green Industries in New Global Economy*. Oxford University Press.

Nanchang Development and Reform Commission. 2010. *Nanchang Low-carbon Pilot City Work Plan Draft (Submitted for Review)* [Nanchang shi di tan shidian gong zuo shishi fangan (bao shen gao)].

"Nanchang's Green and Low-carbon Development Path [Lü se nanchang de ditan fazhan zhi lu]." 2012. *Beijing Weekly* (Beijing Zhoubao), September 11.

Nanchang Housing and Urban–Rural Development Bureau and Nanchang Development and Reform Commission. 2012. *Nanchang Green Building Medium-Long Term Plan (2012–2020).*

Nanchang Housing and Urban–Rural Development Bureau and Nanchang Development and Reform Commission. 2017. *Notice on the Issuance of the Nanchang Green Building Development and Oversight Implementing Regulations (2017–2020)* [Guanyü yinfa Nanchangshi tuijin lüse fazhan guanli gongzuo shishi xize (2017–2020)].

Nanchang Metro. 2021. "The President of Nanchang Metro Li Yunfeng Interviewed by Nanchang Television Station [Zhu tu wu cheng jian she nan chang di tie zhe me zuo shi ren da dai biao li yunfeng jie shou nan chang dian shi tai zhi bo cai fang]." Nanchang: Nanchang Metro.

Nanchang Municipal People's Congress. 2015. *Nanchang Building Market Oversight Regulation.*

Nanchang Municipal People's Congress. 2016. *Regulations on the Promotion of Low-Carbon Development of Nanchang City.*

Nanchang Municipal People's Government. 2011. *Notice on the Issuance of the Nanchang's Low-Carbon Pilot Program Action Plan* [Nanchang shi ren min zheng fu guan yü yin fa nan chang guo jia di tan shi dian gong zuo shi shi fang an de tong zhi].

"National Carbon Emissions Trading Capacity Building Center Settled in Shenzhen [Quan guo tan shi chang neng li jian she zhong xin luo hu shen zhen]." 2016. *China Energy News*

(Zhongguo nengyuan bao), March 28. https://paper.people.com.cn/zgnyb/html/2016-03/28/content_1665161.htm.

National Bureau of Statistics of China. 2024. *Urban Transformation—Report No. 19 in the Series on 75 Years of Economic and Social Development in the People's Republic of China* [Cangsang jübian huan xinyan chegnshi fazhan qi xin pian—Xinzhongguo 75 nian jingji fazhan chengjiu xilie baogaoz hi shijiu], September 23. https://www.stats.gov.cn/zt_18555/ztfx/xzg75njjshfzcj/202409/t20240923_1956630.html.

National Center for Climate Change Strategy and International Cooperation. 2013. *Study on the Current Status and Future Work Directions of Low-carbon Pilot Practices in China* [Woguo ditan shidian shijian xianzhuang ji weilai gongzuo fangxiang yanjiu]. Beijing: National Center for Climate Change Strategy and International Cooperation.

National Center for Climate Change Strategy and International Cooperation. 2016. "National Center for Climate Change Strategy and International Cooperation Releases Research Results of the 'Zhenjiang Index' for Low-Carbon Development [Guojia qihou zhanlue zhongxin ditan fazhan 'zhenjiang zhishu' yanjiu chengguo fabu]." December 8. Beijing: National Center for Climate Change Strategy and International Cooperation.

National Center for Climate Change Strategy and International Cooperation. 2017. *Progress Report of China's Low-Carbon Pilot Provinces and Cities* [Zhongguo ditan shengshi shidian jinzhan baogao]. Zhongguo Jihua Press.

National Development and Reform Commission. 2008. *China's Policies and Actions towards Address Climate Change 2008.*

National Development and Reform Commission. 2014. *The National Development and Reform Commission's Notice on the Assessment Measures for Carbon Intensity Reduction Targets* [Guo jia fa zhan gai ge wei guan yü yin fa dan wei guo nei sheng chan zong zhi er yang hua tan pai fang jiang di mu biao ze ren kao he ping gu ban fa de tong zhi].

National Development and Reform Commission. 2023a. *Notice from the General Office of the National Development and Reform Commission on Issuing the List of the First Batch of Carbon-Peaking Pilots* [Guanyü yinfa shoupi tandafeng shidian mingdan de tongzhi]. https://www.ndrc.gov.cn/xwdt/tzgg/202312/t20231206_1362474.html.

National Development and Reform Commission. 2023b. *National Plan on Carbon-Peaking Pilot Program* [Duojia tandafeng shidian jianshe fangan]. November 15. https://www.ndrc.gov.cn/xxgk/zcfb/tz/202311/P020231106567282826127.pdf.

National Development and Reform Commission and German Corporation for International Cooperation (Deutscher Gesellschaft für Internationale Zusammenarbeit). 2013. *Sino-German Hubei Training Conference on Climate Change Capacity Building Training, Phase 2, Conference Manual* [Zhong de hu bei ying dui qi hou bian hua neng li jian she pei xun hui di er qi ke cheng hui wu shou ce].

North, Douglass. 1990. *Institutions, Institutional Change and Economic Performance.* Cambridge University Press.

O'Brien, Kevin J., and Lianjiang Li. 1999. "Selective Policy Implementation in Rural China." *Comparative Politics* 31, no. 2: 167–186.

OECD (Organization for Economic Co-operation and Development). 2020. *Renewable Energy (indicator).* OECD. https://www.oecd.org/en/data/indicators/renewable-energy.html.

Office of Governor Gavin Newsom. 2023. "What Governor Newsom's Trip to China Accomplished." October 30. https://www.gov.ca.gov/2023/10/30/what-governor-newsoms-trip-to-china-accomplished/.

Oi, Jean C. 1999. *Rural China Takes Off: Institutional Foundations of Economic Reform.* University of California Press.

"One Disciplinary Case Prompts 19 to Voluntarily Return Illicit Funds [Yi ci an jian jing shi, 19 ren zhu dong shang jiao wei ji kuan]." 2019. *Discipline Inspection News* (Zhongguo ji jian jian cha bao), January 16.

Pan, Jiahua, and Guiyang Zhuang. 2018. *Innovation for Low-carbon Development in Xiamen* [Xiamen ditan chengshi chuangxin fazhan yanjiu]. Social Sciences Academic Press.

Pan, Jiahua, Guiyang Zhuang, Shouxian Zhu, and Ying Zhang. 2013. *Reconstruction of China's Low-Carbon City Evaluation Indicator System: A Methodological Guide for Applications.* Social Sciences Academic Press.

Paterson, Matthew. 2021. *In Search of Climate Politics.* Cambridge University Press.

Peng, Yuan, and Xuemei Bai. 2018. "Experimenting towards a Low-Carbon City: Policy Evolution and Nested Structure of Innovation." *Journal of Cleaner Production* 174: 201–212. https://doi.org/10.1016/j.jclepro.2017.10.116.

People's Republic of China. 2023. *The People's Republic of China Fourth National Communication on Climate Change.* https://unfccc.int/sites/default/files/resource/China_NC4_English.pdf.

Perry, Elizabeth J. 2011. "From Mass Campaigns in Managed Campaigns." In *Mao's Invisible Hand: The Political Foundations of Adaptive Governance in China*, edited by Sebastian Heilmann, and Elizabeth J. Perry. Harvard University Asia Center.

Pressman, Jeffrey L., and Aaron Wildavsky. 1984. *Implementation: How Great Expectations in Washington Are Dashed in Oakland; Or, Why It's Amazing That Federal Programs Work at All, This Being a Saga of the Economic Development Administration as Told by Two Sympathetic Observers Who Seek to Build Morals on a Foundation.* 3rd ed. University of California Press.

"Promoting Large-scale Development of Green Buildings in Nanchang, Jiangxi Province [Jiang xi sheng nan chang shi tui dong lü se jian zhu gui mo hua fa zhan ce ji]." 2016. *China Construction News* (Zhong guo jian she bao), March 2.

"Pursuing Carbon Peaking and Carbon Neutrality Actively and Prudently [Ji ji wen to tui jin ta da feng he tan zhong he]." 2023. *Renmin Daily*, April 6.

Qi, Ye, Li Ma, Huanbo Zhang and Huimin Li. 2008. "Translating a Global Issue into Local Priority: China's Local Government Response to Climate Change." *The Journal of Environment & Development* 17, no. 4: 379–400.

Qi, Ye, Qijiao Song, Xiaofan Zhao, Shiyong Qiu, Tom Lindsay, et al. 2020. *China's New Urbanisation Opportunity: A Vision for the 14th Five-Year Plan.* https://urbantransitions.global/en/publication/chinas-new-urbanisation-opportunity-a-vision-for-the-14th-five-year-plan/, London, UK, and Washington, DC: Coalition for Urban Transitions.

Rabe, Barry G. 2004. *Statehouse and Greenhouse: The Emerging Politics of American Climate Change Policy.* Brookings Institution Press.

Ran, Ran. 2013. "Perverse Incentive Structure and Policy Implementation Gap in China's Local Environmental Politics." *Journal of Environmental Policy & Planning* 15, no. 1: 17–39.

Read, Benjamin L. 2018. "Serial Interviews: When and Why to Talk to Someone More than Once." *International Journal of Qualitative Methods* 17: 1–10.

REN21. 2011. *Renewable 2011 Global Status Report.* Paris: REN21.

REN21. 2024. *Renewables 2024 Global Status Report Collection, Global Overview.* https://www.ren21.net/gsr-2024/modules/global_overview. Paris: REN21.

Sabel, Charles F., and Jonathan Zeitlin. 2012. "Experimentalist Governance." In *The Oxford Handbook of Governance*, edited by David Levi-Faur. Oxford University Press.

Sabel, Charles F., and David G. Victor. 2022. *Fixing the Climate: Strategies for an Uncertain World.* Princeton University Press.

Schreurs, Miranda A. 2008. "From the Bottom Up: Local and Subnational Climate Change Politics." *Journal of Environment & Development* 17, no. 4: 343–355.

Schreurs, Miranda A. 2024. "Jockeying for Climate Leadership Amidst Rising Global Tensions: China, USA and the European Union." In *Europe in an Era of US–China Strategic Rivalry: Challenges and Opportunities from an Outside-in Perspective*, edited by Sebastian Biba. Springer.

Schreurs, Miranda A., and Yves Tiberghien. 2010. "European Union Leadership in Climate Change: Mitigation through Multilevel Reinforcement." In *Global Commons, Domestic Decisions: The Comparative Politics of Climate Change*, edited by Kathryn Harrison, and Lisa McIntosh Sundstrom. MIT Press.

Schröder, Miriam. 2012. *Local Climate Governance in China: Hybrid Actors and Market Mechanisms.* Palgrave Macmillan.

Selznick, Philip. 1984. *Leadership in Administration: A Sociological Interpretation.* University of California Press.

"Seven Provinces and Cities to Launch Carbon Emissions Trading Pilots [Guonei qi shengshi jiang qi dong tan jiaoyi shidian]." 2011. *China Energy News*, November 21.

Shen, Shiran Victoria. 2022. *The Political Regulation Wave: A Case of How Local Incentives Systematically Shape Air Quality in China.* Cambridge University Press.

"Shenzhen Carbon Emissions Trading System Has Reduced More Than 58 Million Tonnes of CO2" [Tan shi chang jiao yi liang chao 5800 wan dun, shenzhen jiang ru he zhu li quan guo shi xian tan zhong he]." 2021. *Zhengquanshibao*, March 18.

Shenzhen Carbon Trading Study Tour. 2013. "Learning from the EU ETS and Developing China's Carbon Market: Insights from Three European Countries [Xue xi jie jian eu-ets jing yan yü jian she zhong guo tan pai fang jiao yi ti xi – ou zhou san guo fang tan lu]." *China Opening Journal* 3: 50–63.

Shenzhen City Development Research Center, and Shenzhen Emissions Exchange. 2015. *One-Year Summary Report on the Operational Performance of the Shenzhen Carbon Emissions Trading System.* GIZ China.

Shenzhen Institute of Building Research (Shenzhen jianke yuan). 2013. "Shenzhen International Low-Carbon City, EXPO Centre: Exhibition Plans." Shenzhen: Shenzhen Institute of Building Research.

Shenzhen Municipal Administration for Financial Regulation. 2021. *Shenzhen Green Finance Association Held its First Meeting* [Shenzhenshi lüse jinrong xiehui diyijie hui uan dahui diyici huiyi Chenggong zhaokai]. July 9.

Shenzhen Municipal People's Congress. 2020. *Regulations of Shenzhen Special Economic Zone on Green Finance* [Shenzhen jingjitequ lüse jinrong tiaoli].

Shenzhen Municipal People's Government. 2012. *Shenzhen Low-Carbon Development Medium-and Long-Term Planning (2011–2020).*

Simon, Herbert A. 1983. *Reason in Human Affairs.* Stanford University Press.

Song, Kaihui, Katherine Burley Farr, and Angel Hsu. 2024. "Assessing Subnational Climate Action in G20 Cities and Regions: Progress and Ambition." *One Earth* 7, no. 12: 2189–2203.

Song, Qijiao, Ming Qin, Ruichen Wang, and Ye Qi. 2020. "How Does the Nested Structure Affect Policy Innovation? Empirical Research on China's Low Carbon Pilot Cities." *Energy Policy* 144: 111695.

Song, Qijiao, Tianle Liu, and Ye Qi. 2021. "Policy Innovation in Low Carbon Pilot Cities: Lessons Learned from China." *Urban Climate* 39: 100936. https://doi.org/10.1016/j.uclim.2021.100936.

State Council. 1996. *The Ninth Five-Year Plan for National Economic and Social Development and Long-Range Objectives Through the Year 2010* [Guomin jingji he shehuifazhan jiuwu jihua he 2010 nian yuanjing mubiao gangyao]. https://www.ndrc.gov.cn/fggz/fzzlgh/gjfzgh/200709/P020191029595686994247.pdf.

State Council. 2004. *People's Republic of China's Initial National Communication on Climate Change 2004* [Zhong hua ren min gong he guo qi hou bianhua chu shi guo jia xin xi tong bao].

State Council. 2021. *Action Plan for Carbon Dioxide Peaking Before 2030*. https://english.www.gov.cn/policies/latestreleases/202110/27/content_WS6178a47ec6d0df57f98e3dfb.html.

Statistics Bureau of Nanchang Municipality and Survey Office of the National Bureau of Statistics in Nanchang (Guojia tongjiju Nanchang diaochadui). 2018. *Statistical Communiqué on the 2018 National Economic and Social Development of Nanchang*.

Statistics Bureau of Nanchang Municipality and Survey Office of the National Bureau of Statistics in Nanchang (Guojia tongjiju Nanchang diaochadui). 2019. *Statistical Communiqué on the 2019 National Economic and Social Development of Nanchang*.

Statistics Bureau of Shenzhen Municipality and Survey Office of the National Bureau of Statistics in Shenzhen (Guojia tongjiju Shenzhen diaochadui). 2017. *Shenzhen Statistical Yearbook 2017*. China Statistics Press.

Statistics Bureau of Shenzhen Municipality and Survey Office of the National Bureau of Statistics in Shenzhen (Guojia tongjiju Shenzhen diaochadui). 2019. *Shenzhen Statistical Yearbook 2019*. China Statistics Press.

Statistics Bureau of Shenzhen Municipality and Survey Office of the National Bureau of Statistics in Shenzhen (Guojia tongjiju Shenzhen diaochadui). 2024. *Shenzhen Statistical Bulletin on the 2024 National Economic and Social Development*. https://tjj.sz.gov.cn/gkmlpt/content/11/11264/post_11264245.html#4222.

Statistics Bureau of Xiamen Municipality and Survey Office of the National Bureau of Statistics in Xiamen (Guojia tongji ju xiamen diaochadui). 2019. *Xiamen Statistical Yearbook 2019*. China Statistics Press.

Statistics Bureau of Zhenjiang Municipality and Survey Office of the National Bureau of Statistics in Zhenjiang (Guojia tongji ju zhenjiang diao cha dui). 2011–2019. *Zhenjiang Statistical Yearbook*. 9 Vols. China Statistics Press.

Steinbacher, Karoline, and Michael Pahle. 2015. *Leadership by Diffusion and the German Energiewende*. PIK (Potsdam-Institut für Klimafolgenforschung) Discussion Paper. Potsdam: PIK.

Stern, Rachel E. 2013. *Environmental Litigation in China: A Study in Political Ambivalence*. Cambridge University Press.

Stokes, Leah Cardamore. 2020. *Short Circuiting Policy: Interest Groups and the Battle Over Clean Energy and Climate Policy in the American States*. Oxford University Press.

Sun, Yixian. 2022. *Certifying China: The Rise and Limits of Transnational Sustainability Governance in Emerging Economies*. MIT Press.

Teets, Jessica C. 2015. "The Politics of Innovation in China: Local Officials as Policy Entrepreneurs." *Issues & Studies* 52, no. 2: 79–109.

Teets, Jessica C. 2018. "The Power of Policy Networks in Authoritarian Regimes: Changing Environmental Policy in China." *Governance* 52, no. 2: 125–141.

Teets, Jessica C., and William Hurst. 2015. "Introduction: The Politics and Patterns of Policy Diffusion in China." In *Local Governance Innovation in China: Experimentation, Diffusion and Defiance*, edited by Jessica C. Teets and William Hurst. Routledge.

Teets, Jessica C., and Nele Noesselt. 2020. "The State of the Field for Governance and Policy Innovation in China." *Journal of Chinese Governance* 5, no. 4: 413–418. https://doi.org/10.1080/23812346.2020.1799641.

Teng, Fei, and Pu Wang. 2021. "The Evolution of Climate Governance in China: Drivers, Features, and Effectiveness." *Environmental Politics* 30, no. Sup. 1: 141–161. https://doi.org/10.1080/09644016.2021.1985221.

Tian, Chuanhao. 2018. *Land Finance in China: History, Reality and Possible Changes* [Zhong guo tu di cai zheng: li shi, xian shi ji ke neng de bian ge]. Zhejiang University Press.

Tie, Ming, Ming Qin, Qijiao Song, and Ye Qi. 2020. "Why Does the Behavior of Local Government Leaders in Low-carbon City Pilots Influence Policy Innovation?" *Resources, Conservation & Recycling* 152: 104483. https://doi.org/10.1016/j.resconrec.2019.104483.

Torney, Diarmuid. 2015. *European Climate Leadership in Question: Policies toward China and India.* MIT Press.

Tsai, Kellee S. 2004. *Back-Alley Banking: Private Entrepreneurs in China.* Cornell University Press.

Tucker, Robert C. 1995. *Politics and Leadership.* Revised ed. University of Missouri Press.

United Nations Environmental Programme (UNEP). 2019. "NBS Good Practices from Chinese Government." May 2019 Carbon Neutral Project for BRICS Leaders Xiamen Meeting. United Nations Environmental Programme (UNEP).

United Nations Environmental Programme (UNEP). 2020. "Xiamen Marathon's Race towards Sustainability." January 16. United Nations Environmental Programme (UNEP).

U.S. Department of State. 2023. *Sunnylands Statement on Enhancing Cooperation to Address the Climate Crisis.* https://2021-2025.state.gov/sunnylands-statement-on-enhancing-cooperation-to-address-the-climate-crisis/.

U.S. Energy Information Administration. 2014. "China Is Now the World's Largest Net Importer of Petroleum and Other Liquid Fuels." March. https://www.eia.gov/todayinenergy/detail.php?id=15531.

van der Kamp, Denise. 2023. *Clean Air at What Cost?The Rise of Blunt Force Regulation in China.* Cambridge University Press.

van Rooij, Benjamin. 2006. "Implementation of Chinese Environmental Law: Regular Enforcement and Political Campaigns." *Development and Change* 37, no. 1: 57–74.

Vogel, David. 1997. "Trading Up and Governing Across: Transnational Governance and Environmental Protection." *Journal of European Public Policy* 4, no. 4: 556–571. https://doi.org/10.1080/135017697344064.

Vogel, David. 2018. *California Greenin': How the Golden State Became an Environmental Leader.* Princeton University Press.

Wang, Alex. 2013. "The Search for Sustainable Legitimacy: Environmental Law and Bureaucracy in China." *Harvard Environmental Law Review* 37: 365–440.

Wang, Alex. 2018. "Symbolic Legitimacy and Chinese Environmental Reform." *Environmental Law* 48: 699–760.

Wang, Alex. 2020. *Banning Coal—A Modest Proposal For U.S.–China Climate Change Cooperation.* Working Paper for the Penn Project on the Future of U.S.–China Relations. https://papers.ssrn.com/sol3/papers.cfm?abstract_id=3717304.

Wang, Dong, Matin de Jong, Jiangbo Bie et al. 2012. "New Explorations in Shenzhen's Low-Carbon Development: The Shenzhen–The Netherlands (Europe) Low-Carbon City Project [Shenzhen di tan fa zhan de xin tan suo: shenzhen zhong he di tan cheng xiang mu jian she jie shao]." In *Shenzhen Economic Development Report (2012)*, edited by Zhong Wu, Hongjuan Liu, and Xiaoyuan Dong. Social Sciences Academic Press.

"Wang Menghui: Build Large-Capacity Express Transit to Advance Xiamen's Island-to-Bay Transition [Wang Menghui: Dazao dayunliang kuaisu jiaotong tuijin haidao dao haiwan zhuanxing]." 2011. *Xiamen TV News*, December 31.

Wang, Shaoda, and David Y. Yang. 2021. *Policy Experimentation in China: The Political Economy of Policy Learning.* Working Paper Series. National Bureau of Economic Research. https://doi.org/10.3386/w29402.

Wang, Yufei, Qijiao Song, Jijiang He, and Ye Qi. 2015. "Developing Low-Carbon Cities through Pilots." *Climate Policy* 15, no. S1: S81–S103. https://doi.org/10.1080/14693062.2015.1050347.

We Are Still In. 2018. "We Are Still in" Declaration. We Are Still In. https://www.wearestillin.com/we-are-still-declaration.

World Bank. 2008. *State and the Trend of Carbon Market*. World Bank. https://openknowledge.worldbank.org/entities/publication/16141b3c-0156-5da4-8d33-cbdd361dcf15.

World Bank. 2012. *Project Information Document (PID) Concept Stage*. World Bank.

World Bank and the Development Research Center of the State Council (the People's Republic of China). 2014. "China's Urbanization and Land: A Framework for Reform." In *Urban China: Toward Efficient, Inclusive, and Sustainable Urbanization*, edited by The World Bank and the Development Research Center of the State Council (the People's Republic of China). World Bank. https://www.worldbank.org/content/dam/Worldbank/document/EAP/China/WEB-Urban-China.pdf.

World Bank. "The World Bank in China." 2024. World Bank. https://www.worldbank.org/en/country/china.

World Low-Carbon and Eco-Economy Conference. 2009. *World Low-Carbon and Eco-Economy Conference Report 9*. Nanchang: World Low-Carbon and Eco-Economy Conference.

White House. 2025. *Protecting American Energy from State Overreach*. April. https://www.whitehouse.gov/presidential-actions/2025/04/protecting-american-energy-from-state-overreach/.

Wu, Fengshi. 2013. "Environmental Activism in Provincial China." *Journal of Environmental Policy & Planning* 15 no. 1: 89–108. doi:10.1080/1523908X.2013.763634.

Wu, Kang. 2014. "China's Energy Security: Oil and Gas." *Energy Policy* 73: 4–11.

"Xi Jinping Presides over the 36th Collective Study of the Politburo of the CPC Central Committee and Delivers an Important Speech [Xi Jinping zhuchi Zhonggong Zhongyang Zhengzhiju di sanshiliuci jitixuexi bing fabiao zhongyao jianghua]." 2022. Xinhua News Agency, January 25. https://www.gov.cn/xinwen/2022-01/25/content_5670359.htm.

Xiamen Academy of Urban Planning and Design. 2017. *Beautiful Xiamen Strategic Plan*. http://www.cacp.org.cn/ghqy/5823.jhtml.

Xiamen Construction Bureau. 2012. *Xiamen Low-Carbon City Construction* [Xiamen di tan cheng shi jian she]. White paper. Xiamen Construction Bureau.

Xiamen Development and Reform Commission. 2017. *Xiamen Development and Reform Commission Report on 2017 Climate Action Outcomes and 2018 Work Plan* [Xiamen shi fagaiwei guanyü yingdui qihoubianhua 2017 nian gongzuochegnxiao he 2018 nian gongzuo shexiang de baogao].

Xiamen Development Research Center (Xiamen shi fazhan yanjiu zhongxin). 2018. *Report on the Operation of Xiamen Low-Carbon Emissions Management Platform* [Xiamen shi tanpaifang zhineng guanli yun pingtai xiangmu zongjie baogao].

"Xiamen Launches Low-Carbon Pilot Demonstration [Xiamen kai zhan di tan shi dian shifan chuang jian]." 2023. *China Environmental News*, October 25. http://epaper.cenews.com.cn/html/1/2023-10/25/02B/2023102502B_pdf.

"Xiamen Low-Carbon Economy Report: Building a Low-Carbon City and Promoting on Low-Carbon Living [Xiamen di tan jing ji bao gao: jian di tan cheng shi zhu zhong di tan sheng huo]." 2012. *China Environmental News* (Zhongguo huan jing bao), August 22.

Xiamen Municipal Government. 2011. *The Outline of the Twelfth Five-Year Plan on Xiamen Economic and Social Development* [Xiamen shi guo min jing ji he she hui fa zhan di shi er ge wu nian gui hua gang yao].

Xiamen Municipal Government. 2014. *Xiamen Green Building Action Implementation Plan*.

"Xiamen Releases the 'Beautiful Xiamen Strategic Plan' [Meili Xiamen zhanlüe quanwen fabu]." 2014. *Xiamen Daily* (Xiamen ribao), March 5.

"Xie Zhenhua: The 1 + N Policy System for Carbon Peaking and Carbon Neutrality to be Released Soon, Aiming to Accelerate Transformation Across Ten Key Areas [Xie Zhenhua:

tan da feng zhong he 1+N zheng ce ti xi ji jiang fa bu shi ge ling yü jia su zhuan xing]." 2021. *Beijing News* (Xinjingbao), June 24.

Xiong, Jihai, Xiping Xi, and Min Fan. 2020. *Low-Carbon City Development: The Case of Nanchang*. China Science Publishing & Media.

Xu, Huaqing, Xuechun Wang, Danyu Tian, Zeyu Zhou, and Xiu Yang. 2017. *Field Research and Summary Report on China's National Low-Carbon Pilot Provinces and Cities* [Guo jia di tan sheng shi shi dian gong zuo diao yan yü zong jie bao gao]. Beijing: National Center for Climate Change Strategy and International Cooperation.

Xu, Zongheng. 2009. *Report on the Work of the Government of Shenzhen Municipality 2009*. Shenzhen Municipal People's Government. https://www.sz.gov.cn/zfgb/2009/gb644/content/post_4992976.html.

Xue, Lan, Udo E. Simons, and Daniel J. Dudek. 2007. "Environmental Governance for China: Major Recommendations of a Task Force." *Environmental Politics* 16, no. 4: 669–676.

Xue, Lan, and Jing Zhao. 2018. "Truncated Decision Making and Deliberative Implementation: A Time-Based Policy Process Model for Transitional China." *Policy Studies Journal* 48, no. 2: 298–326. https://doi.org/10.1111/psj.12290.

Yang, Shiming. 2023. "Growing Apart: China and India at the Kigali Amendment to the Montreal Protocol." *Global Environmental Politics* 23, no. 2: 74–101. https://doi.org/10.1162/glep_a_00698.

Yang, Xiu, Xue-Chun Wang, and Ze-yu Zou. 2018. "Development Path of Chinese Low-Carbon Cities Based on Index Evaluation." *Advances in Climate Change Research* 9: 144–153.

Yang, Xuedong. 2011. "China Local Government Reform in the Past Decade: An Evaluation Based on Chinese Local Governance Innovations Awards [Guo qü 10 nian de zhong guo di fang zheng fu gai ge ji yü zhong guo di fang chuang xin jiang de ping jia]." *Journal of Public Management* 1: 81–93.

Yasuda, John Kojiro. 2015. "Why Food Safety Fails in China: The Politics of Scale." *China Quarterly* 223: 745–769. https://doi.org/10.1017/S030574101500079X.

Yasuda, John Kojiro. 2024. "Explaining Policy Failure in China." *China Quarterly* 257: 3–19. https://doi.org/10.1017/S0305741023000711.

Yu, Xuan, Manhong Shen, Di Wang, and Bernadette Tadala Imwa. 2019. "Does the Low-Carbon Pilot Initiative Reduce Carbon Emissions? Evidence from the Application of the Synthetic Control Method in Guangdong Province." *Sustainability* 11, no. 14: 3979. https://doi.org/10.3390/su11143979.

Zhao, Hui, Zhu Xufeng, and Ye Qi. 2016. "Fostering Local Entrepreneurship through Regional Environmental Pilot Schemes: The Low-Carbon Development Path of China." *China: An International Journal* 14, no. 3: 107–130.

Zhang, Yefei. 2019. *2019 Zhenjiang Municipal Government Work Report* [Zhenjiang shi zheng fu gong zuo bao gao 2019]. January 9.

Zhang, Xuefan, and Yanling He. 2024. "Bureaucratic Entrepreneurship: How Frontline Bureaucrats Promote Policy Innovation." *Journal of Public Policy* 44, no. 2: 411–435. https://doi.org/10.1017/S0143814X24000023.

Zhang Xuejin. 2015. Ministry of Water Resources Treating the Development of Small Hydropower with an Objective Attitude [Shuilibu Zhang Xuejin: yongkeguantaidukandai xiaoshuidian fazhan]." *China Power News* (Zhongguo dianli bao), July 17, 2015.

Zhenjiang Development and Reform Commission. 2017a. *Report on Zhenjiang's Low-carbon City Development* [Zhenjiang ditan chengshi jianshe qingkuang huibao].

Zhenjiang Development and Reform Commission. 2017b. *Summary of the International Low Carbon Conference, Zhenjiang* [Guoji ditan Zhenjiang dahui qingkuang zongjie].

Zhenjiang Development and Reform Commission. 2018. *Jiangsu Provincial Standard: Preparation Instructions for the Low-Carbon City Evaluation System* [Jiang su sheng di fang biao zhun "di tan cheng shi ping jia zhi biao ti xi" bian zhi shuo ming]. September 27.

Zhenjiang Housing and Urban–Rural Development Bureau (Zhenjiang zhufang yü chengxiang jianshe weiyuanhui). 2018. *Zhenjiang Green Building Demonstration City Implementation Report* [Zhenjiang shi lü se jian zhu shi fan cheng shi shi shi zong jie bao gao].

Zhenjiang Institutional Organization Bureau (Zhenjiang shi jigou bianzhi weiyuanhui). 2017. *Notice on the Establishment of the Low-Carbon Development Division at the Zhenjiang Development and Reform Commission* [Guan yü zai shi fa gai wei she li di tan fa zhan chu de tong zhi]. December 25.

Zhenjiang Municipal Government. 2020. *Zhenjiang Releases Nine Major Action Tasks for Low-Carbon Development* [Woshi di tan jianshe jiüda xingdong renwu fabu]. July 2.

"Zhenjiang Tells the World a Low-Carbon Story [Zhenjiang: si fu lü se zhi yue xiang shi jie jiang hao di tan gu shi]." 2019. *Yangtz Evening Post*, October 24.

Zhou, Li-an. 2008. *Local Government in Transformation: Bureaucratic Incentives and Governance* [Zhuan xing zhong de di fang zheng fu: guan yuan ji li yü zhi li]. Truth & Wisdom Press.

Zhou, Nan, Gang He, and Christopher Williams. 2012. *China's Development of Low-Carbon Eco-Cities and Associated Indicator Systems.* Ernest Orlando Lawrence Berkeley National Laboratory.

Zhu, Yapeng, and Diwen Xiao. 2015. "Policy Entrepreneur and Social Policy Innovation in China." *Journal of Chinese Sociology* 2, no. 10: 56–76. https://doi.org/10.1186/s40711-015-0012-z.

Zhuang, Guiyang. 2021. *Theory, Method and Practice of Low Carbon City* [Ditan chegnshi de lilun fangfa yü shijian]. China Social Sciences Press.

Index

For the benefit of digital users, indexed terms that span two pages (e.g., 52–53) may, on occasion, appear on only one of those pages.
Tables, figures, and boxes are indicated by an italic *t*, *f*, or *b*.